RAISING
LESS CORN,
MORE HELL

RAISING
LESS CORN,
MORE HELL

The Case for
the Independent Farm
and Against
Industrial Food

GEORGE PYLE

PublicAffairs
New York

BOOK DESIGN BY JANE RAESE
Text set in 11 point Chaparral

Library of Congress Cataloging-in-Publication Data
Pyle, George.
Raising less corn, more hell: the case for the independent farm
and against industrial food / George Pyle.
p. cm.
ISBN 1-58648-115-0
1. Farms, Small—United States. 2. Family farms—United States.
3. Farms—United States. 4. Agricultural industries—United States.
5. Farm corporations—United States. 6. Agricultural and state—
United States. 7. Agricultural—Economic aspects—United States.
8. Agricultural—Environmental aspects—United States.
I. Title.
HD1476.U5P95   2005
338.1'0973—dc22
2005041902

FIRST EDITION

2 4 6 8 10 9 7 5 3 1

IN MEMORY OF MY FATHER

Contents

CONTENTS

PART THREE: SECURITY

Burn down your cities and leave our farms, and your cities will spring up again as if by magic, but destroy our farms and the grass will grow in the streets of every city in the country.

—William Jennings Bryan, 1860–1925

Farmers should be raising less corn and more hell!

—Mary Elizabeth Lease, 1853–1933

Prologue: Searching for Roots
Or, How I Learned to Start Worrying and Love the Small Farm

The earth is not dying, it is being killed.
And those who are killing it have names and addresses.

—Utah Phillips

This book began as a great many blank pages.

That is true of all books, of course. But this one more than most.

I wrote for Kansas newspapers for twenty-three years. That's a lot of blank pages and, later, blank computer screens to fill. And, in Kansas, it is hard to do that without being called upon, from time to time, to write about agriculture—farming, food production, and all the social, political, and economic issues that go with the subject. Like most reporters on a daily deadline, I explained more than I understood.

In the beginning, I filled many pages with information that intermittently issued from any number of government officials, university experts, agribusiness spokesmen, and, when farmers were having especially hard times, even one or two of them. Like most reporters writing for small dailies, I wrote about agriculture only when it worked in among many other assignments, from murder trials to first-day-of-school features. The "farm beat," for reporters who covered it, consisted mostly of tracking the highs and lows of market prices; noting, without comment, the introduction of a new machine, chemical, or government program; and snapping the firing-squad-style photo of a farm family that had won the local Soil Conservation District award.

As a reporter, I gave little thought to agriculture as a subject, in part because I was, by Kansas standards anyway, a city kid who wanted to write about goings-on in the halls of power. Like many young people from the Great Plains, I harbored a vague feeling that farmers and farming were somehow leftovers from a more primitive age and, as a farm boy named Luke Skywalker once said about his boyhood home, about as far away from the bright center of the universe as anyone could get.

My father was raised on a Kansas farm and recounted for me tales of milking an infinite number of cows and riding a horse to school. But because of physical handicaps brought on by a childhood attack of polio, he had to find another way to make a living. A university scholarship offered to handicapped people—delivered to his farmhouse by a one-armed man—gave him the means to leave farming. To this day, though, I have relatives who make all or part of their living by farming. I always knew they worked very hard indeed. I knew that food came from farms, not the grocery store. But there never seemed to be any problem getting it from the farm to the store. It didn't cost very much. It seldom made anyone sick. As a cub reporter, I didn't see much of a story there.

Besides, covering agriculture felt like covering a glacier. People knew the glacier was moving, apparently inexorably, in a known direction. And inevitability is not newsworthy.

Farming, everyone thought, was shifting away from Thomas Jefferson's ideal of the independent yeoman farmer, the backbone of democracy, and toward more of a Henry Ford or, more precisely, Monsanto model, with the many unfortunate bump-kins condemned to scratching a meager living out of the earth being replaced by fewer and fewer mechanized, digitized, capital-ized, genetically enhanced agribusinesspersons.

As with the motion of a glacier, the only newsworthy event was when a large chunk of it fell off, which occurred figuratively whenever there was a rash of farm bankruptcies and/or a multi-billion-dollar emergency aid package rushed through Congress. Occasionally, the glacier might even seem to move backward, as when the 1972 Russian Wheat Deal caused a spike in grain prices, which made Great Plains farmers very happy and, on paper, rich.

But truly frozen in place was any critical thinking, not only by a great many farmers and farm experts but also by both the main-stream media and what passes then and now for a farm press.

Now, nearly thirty years later, it is time to tell the story I could not see before: The industrialization of agriculture is not in-evitable. It is a monumentally bad idea. And it is based on a lie.

The bad idea is the increasing concentration—economic, politi-cal, and genetic—of the ways in which our food is produced. It is a clear and present danger to the economies of both developed and developing nations. It is the single most pressing issue facing our environment. And it is a national-security—or, to use the post–9/11 term, homeland-security—issue of the first magnitude.

Here is the lie: The world is either short of food or risks being short of food in the near future.

The truth is that the problems of food, for those who grow it and those who need it, are almost always problems of overpro-duction, not underproduction. Yet the phantom of underproduc-tion haunts almost all agricultural policy decisions. Nations with large numbers of hungry people are almost always nations with hardworking farmers who cannot compete with the government-subsidized staple crops grown in massive quantities in the United States and the nations of the European Union, then dumped on the market at prices below the cost of production. Local and regional shortages of food are also more properly blamed on corrupt governments, war, and the industrialization of Third World agriculture. All of these factors have displaced locally adapted, self-supporting farm communities in favor of giant plantations that export commodities such as coffee, bananas, sugar, and cocoa to the United States and Europe and dump the surplus population into disease-ridden urban slums.

. . .

As a newspaper summer intern in 1977, I spent a good portion of my time writing articles about the annual wheat harvest. My job—which the veteran journalists didn't want to be bothered with—was to call around to grain elevators that were receiving loads of freshly cut wheat and to the temporary offices set up by the Kansas Department of Human Resources to match ripe wheat fields with custom cutters (threshing crews), gangs of workers who traveled from Texas to Canada every summer towing massive combines for cutting the wheat as it matured from south to north. I dutifully recorded the progress of the harvest, from test cutting to full-scale reaping. I noted reports of bushels per acre, test weights, and moisture content. I more or less grasped, without much explanation sought or given, that the first two numbers should be high, and the third one should be low. I don't recall including the prices farmers received for their year's labor and expense. Everybody looked to the small print on the newspaper's business page—or, in Kansas, the farm/business page—for that.

My editors that summer pronounced my harvest work adequate—accurate but uninspired—certainly much less interesting than my coverage of the small-town bank robbery or my lengthy how-to article on selecting the correct bicycle. Harvest was the single most important week of the economic year in that part of the world, but the newspaper coverage, usually relegated to rookies or interns, resembled the way we fielded calls from high school football coaches and filled out box scores (except that I used complete sentences).

It is doubtful anyone found my daily articles on the progress of the harvest all that useful. Farmers got their important data— shifts in market prices, availability of temporary labor—from farm radio, county extension offices, and word of mouth. City folks were likely as puzzled by the significance of moisture-content figures as I was, if they ever bothered to read that far into the articles.

As I moved up the journalistic food chain and became the editorial voice of a relatively large newspaper in the center of Kansas, it became my pleasure and my burden to have and express opinions. No longer could I merely relay the properly attributed views and prejudices of others. Now I had to know not only what was happening but also what should be happening, what shape national and even world farm policy should take, and how much, or how little, support farmers were owed from the taxpayers.

That is when the pages became increasingly blank, often painfully so.

Issues that were important to a significant portion of my readers included farm subsidies, pesticide restrictions, the availability of foreign markets, and the use of American food supplies as a weapon—to be withheld from nations that did not meet our standards of behavior. These matters even affected people who were three or four economic steps removed from the farm and the grain elevator and who thus did not realize their importance.

Into the 1980s and 1990s, the industrialization of agriculture (though it wasn't always called that) became a crucial issue. Kansas, like other states, was seeing fewer farm families and more large operations, and the family operations that did survive were increasingly squeezed between a shrinking number of suppliers for expensive seed, fertilizer, and fuel on one end, and a diminishing number of buyers for their wheat, corn, cattle, and pigs on the other.

Nationally, the number of farmers continued to fall, from its peak of more than 30 million in the 1930s to barely 2 million in 1995. Of more political significance, the percentage of Americans who farmed for a living tumbled from 42 percent at the beginning of the twentieth century to a mere 1.6 percent at the end of it.

Cows, pigs, and, especially, chickens were owned by, or under contract to, huge processing companies, which built giant, job-creating, water-hogging, and, in the view of many, community-destroying protein factories in previously preindustrial towns. Even the environmental aspects of traditional farming—the smell, the dust, the runoff from fields and feedlots—became pressing problems as sprawling suburban areas grew more populated, more politically powerful, and, in the farmers' view, more cluelessly intolerant of the amount of blood, sweat, and fertilizer necessary to provide city folks with their daily bread, and the meat to go on it.

For a Kansas newspaper editor to have no opinion on farm issues would be akin to a Florida counterpart having no thoughts on Medicare or a French editor expressing no opinions on wine or cheese. For me, with increasing frequency and frustration, pages that should have been filled with insightful opinions and expressions of alarm over the woes of Kansas farmers remained blank, to be replaced with discussion of education, taxes, highways, public safety, and civil rights. I knew more about such topics because of my long reporting career, which had not, however, informed me on the issues of farming.

Most of my encounters with farmers were still limited to individuals who also held some public office. These included the county commissioner who introduced me to the term "toad-strangler"—to denote a heavy rain—and believed, along with many other farmers and agricultural experts, that the farm crisis of the 1980s had served mostly to weed the bad farmers out of the business. Many farmers and academics alike saw consolidation and industrialization as inevitable, if not downright positive. Why should agriculture be any different from auto manufacturing or any other segment of the modern, globalized economy?

The fatalism in the glacier story—"the inevitable is not newsworthy"—dominated the mood of farming, and of what passes for farm journalism, through much of the 1980s and 1990s. There were occasional outbursts of concern and anger, such as the "tractorcades" to Washington, D.C., and the Farm Aid concerts. But the mood was largely one of resignation.

Farming was becoming less of an art and more of a business. Get big, the pundits said, or get out. Large farm organizations such as the American Farm Bureau Federation urged their members to align themselves with the big processors, such as ConAgra, Tyson, or Archer Daniels Midland, to have the best chance of garnering at least a piece of the agribusiness pie. Any suggestion that government step in and protect farmers from the predatory practices of the agribusiness giants was (and is) usually rejected by these so-called mainstream farm groups on the grounds that the free market would provide the necessary remedies. Thus, the

United States was the only nation on Earth to pretend that agriculture worked in a free market.

By the late 1990s, I was opening an increasing number of new computer files intending to take an editorial stand on some aspect of farm policy. But likewise growing was the number of files I killed without following through on the subject.

My personal sympathy lay with the farmers. I am descended from farmers, after all. And I lived in Kansas, where sunflowers and wheat are official state emblems. Where college football fans don't do the wave but rather stand and sway upraised arms side to side like tall stalks of wheat waving in the wind. I graduated from Wichita State University, which for the past fifty years has been big on aviation engineering and urban and minority studies, but whose mascot remains an anthropomorphic bundle of wheat called the Wheatshocker. The increasingly urban nature of the school moved everyone to shorten the nickname to "Shockers," and the mascot became the "WuShock." But students know that the nickname came from those summers a great many years ago when members of the football team stayed in shape by "shocking" wheat—beating the stalks to separate the grain from the rest of the plant. One of my college roommates, a journalist named Tim Pouncey, wrote that Wheatshockers should be glad those bygone athletes did not choose to spend their summer cleaning out stables.

Sympathies alone are no basis for good editorial policy, much less wise public policy. If my newspaper was to call for government action to preserve the small, independent farm, or if it was to recognize not only the inevitability but also the wisdom of the marketplace and the increasing consolidation of the food-production system, I would have to learn more about agriculture, particularly the politics and economics of it. I needed to learn what was happening, why, and if it was good or bad. And given the rate at which the big was crowding out the small, I sensed I needed to do all that soon, before it was too late to recommend a change in course.

A golden opportunity to do just that came not from a one-armed man but from the Society of Professional Journalists in the form of the Eugene Pulliam Fellowship for Editorial Writers. That support gave me means to travel, gather materials, and take time away from the tyranny of the daily deadline, all necessary factors for me to get some handle on this troubling and complex subject. A further opportunity, although it didn't seem golden at the time, came when the newspaper chain I had worked for twenty-three years decided that a full-time opinion writer, especially one who kept wanting time off to write a book, was more than it could justify to its stockholders. That separation, which I never would have been brave enough to initiate, led me to the Land Institute, a natural-systems agriculture research organization in the middle of Kansas, where I became a founding director of the Prairie Writers Circle and where I had still more time and support to continue the research and writing that culminated in this book.

$$\cdot \quad \cdot \quad \cdot$$

The pattern of change in agriculture—the loss of small farmers, the concentration of ownership, the industrialization of meat production, the creation of a chemical-dependent culture of grains, the urge to monkey around with the genetic components of food—is not inevitable, and it is not good. The shift, therefore, is newsworthy, much more so than the mainstream press, or even media aimed at farmers, might lead one to believe.

Modern trends and theories in agriculture are depopulating the countryside, spoiling the land, squandering the water, poisoning the food, deepening the global divisions between rich and poor, and threatening whole ecosystems. These trends are promoted by the greed of agribusiness giants, aided and abetted by government. It is allowed by consumers and voters who, if they think about it at all, have bought the common, sincerely held, and utterly wrong assumptions that only tons of pesticides and fertilizers stand between us and famine, that food is no different from any other consumer product in its ability to be industrialized, and that decisions leading people to move off the farm and into the city—whether in Illinois, India, or Ivory Coast—are doing everyone a favor. In order to keep these beliefs widespread in the culture, American democracy is being undermined, as government, industry, academia, and farm organizations treat the citizens of this and other nations as if they were so many mushrooms—keeping people in the dark and covered with manure.

When it appears that a particular change will occur whether one likes it or not, the tendency is often to learn to like it. As then–Secretary of Agriculture Ann Veneman happily told a Kansas television reporter in 2001, "There are many opportunities for producers to become part of the supply chain of the big companies."

"In other words," retorted Kansas cattleman Mike Callicrate on his website, www.nobull.net, "opportunities to become indentured slaves for her agribusiness bosses."

Veneman and Callicrate represent the extremes of thinking about American agriculture. Veneman, like many other government officials whose real job is to promote rather than regulate food production, could see nothing wrong with ever-increasing centralization and industrialization of agriculture. Boosters see this pattern as nothing more than the same kind of mass-production efficiency that not only fostered the invention of automobiles, televisions, and computers but also made them cheap enough for mass consumption. Callicrate, on the other hand, views the industrialization of agriculture as a conspiracy. Not a hidden conspiracy of the sort imagined to be behind, say, the death of President Kennedy, but an open, straightforward collusion of big business, government, and even some academics to empower a few giant agribusiness firms and transform once-proud American farmers into serfs.

The majority of farmers and those who work with them probably agree, deep down, with Callicrate's theory that agribusiness hegemony and government policy are crowding them out of business and off the land. They certainly see that Callicrate is more realistic than are Veneman and their local Farm Bureau. Farmers know how many of their old friends have had to sell out and move to town, how many of their children went off to college, maybe even to agricultural college, never to return. They know machinery, seed, fertilizer, and chemicals cost more but the elevator pays less for their corn and wheat. They know that the old-time livestock auction, which drew enough bidders to boost the price of cattle and hogs, at least a little, has been replaced almost completely by the single-buyer contract. They know that if it were not for billions upon billions in government aid, much more deeply hated by the few who receive it than by the many who provide it, a great many more of them would have been out of business long ago.

Farmers know that the increased power of big agribusiness leaves them despondent, dependent, and in debt. They know that big agribusiness wants it that way so that the farmers, stubbornly clinging to ancestral property and the shards of a cherished way of life, will hang on as long as they can, go into debt, and take on a second job in town, in order to provide the great black hole of food processing with raw materials cheaper than what could be obtained in any other set of circumstances. They know that even the large government subsidies they receive are not for them to keep but to pass on—to launder—to the agribusiness concerns that sell them supplies and equipment.

Cynical, hopeful, or utterly realistic, many farmers cling to the belief that at least some of them will remain because the agribusiness giants that draw life from the farmers cannot afford to kill them off once and for all. The giants know better than to get into the business of actually raising cows, pigs, corn, and soybeans. Too much risk. Too little reward.

Why own the farm when you can own the farmer?

The lies and confusion surrounding agricultural issues are buttressed by the widely held belief that any effort to slow the trend toward industrialization is mere fuzzy-headed nostalgia. Any suggestions that governmental and corporate actions that have actively undermined the independent farm are wrong or should be reversed are attacked as soft-hearted sympathy for Auntie Em and Uncle Henry. Such malcontents, the argument goes, instead should realistically accept the need for and benefits of mass-production agribusiness. For this reason, I generally avoid the term "family farm," even as I describe and advocate the small-scale, independent agriculture that resembles, well, a family farm.

Saving the small, independent farm is not something that should be done for the benefit of small, independent farmers. Independent farmers, despite their place in the American pantheon, do not automatically deserve to be preserved by policy choices any more than did the Pony Express or the town crier. But, like many things done to gladden the bleeding heart, every step taken to allow the independent farm to survive into the twenty-first century will not be an act of charity or nostalgia but an act of self-preservation for all of us, even for people who never have and likely never will set foot on a working farm.

The fact is that government policy choices and corporate profiteering, not the wisdom of the marketplace, have undermined the independent farm and concentrated ownership of the very stuff of life into fewer and fewer hands. This despite the fact that true free-market considerations of efficiency and quality have been, and still are, better served by a great many independent, entrepreneurial, and even idiosyncratic independent farmers who lovingly watch the land, care for the soil, husband their resources, and treat the earth so that it will consent to feed everyone not only now but for many generations to come.

The economic, geographic, and genetic concentration of the plants and animals that provide our food do not serve the farmer, and more important, they do not serve the consumer or the nation. This concentration makes the food supply more vulnerable to attack, both by natural pests and by human terrorists. It leaves Americans dependent on foreign-owned or transnational corporations for their very lives. It undermines a sustainable system of food production, which is, inevitably, still the foundation of any economic system in the First, Second, or Third World. Concentration undermines democracy and seeks to deny parents the most basic knowledge about what they are feeding their children. It traps poor nations in cycles of poverty and dependence by undercutting the value of their agricultural production. Most important, it is the most pressing and most threatening aspect of a mind-set that believes humanity must conquer Nature, by any means necessary, rather than learn to live with it and by its laws.

The independent farmer is what environmentalists call an

"indicator species"—necessary and misunderstood for the same reasons that people often miss the significance of the spotted owl or the snail darter. Like those creatures, independent farmers are not to be saved for their sake alone but as proof that the overall ecosystem remains healthy. The point is not to preserve independent farmers because they are cute and cuddly, like pandas, or even the object of blood sport, like pheasants. The point is, or should be, to keep independent farmers because if they thrive, we know the world is geared to allow that institution. And if the world is geared for the survival of farmers, only then can we have hope that they will be able to support the rest of us.

Furthermore, like the spotted owl and the snail darter, the endangered species of the independent farmer cannot rescue itself. In the same way a nest of owls might be hidden away in a remote stand of trees or even an abandoned human structure, some individual farmers will be able to survive, even thrive, by escaping into clever niches such as organic farming or direct marketing to increasingly choosy consumers. But as a group, independent, land-nurturing sons and daughters of the soil are powerless to oppose the forces of consolidation on their own. The problem is not that they are not creative, clever, or emotionally strong but rather that, at a mere 2 percent of the American population, there aren't enough of them. And that is why, as has been pointed out to me, there are not very many individual farmers discussed in this book.

In the late nineteenth century, a Kansas populist rabble-rouser named Mary Elizabeth Lease told farmers that if they were to hold their own with the banks, railroads, and meatpackers, they should be "raising less corn and more hell." That likely was a phrase drawn from not only the price-depressing nature of farm surpluses at the time but also the need for farmers to devote less time to traditional efforts in the field and take on the task of organizing millions to face down the growing bear of giant agribusiness. In the early twenty-first century, we have come to a point where if the independent farm—and all that it has to offer our bodies, our wallets, and our world—is to survive, the consumer and the voter will have to raise more hell—a lot more. We will have to vote with our food-buying dollar, and we will have to outbid the agribusiness corporations for the attention of our lawmakers. We will have to begin demanding that everything from the regulation of pesticides to the pricing of milk be set with an eye toward the long-term health of consumers, here and around the world, not the short-term needs of the industrialized companies.

Henry David Thoreau said, "In nature is the salvation of the world." And in the independent farmer, in his unique ability to care for the land, in her need to protect her local environment for the sake of her children, is the salvation of Nature.

PART ONE

WEALTH

STALIN'S REVENGE

> Efficient markets by definition have
> so many participants that no single one
> can affect the market price.
>
> —George Soros

When the Cold War memorial is built on the Mall in Washington, D.C., space should be left, alongside the depictions of the soldiers and the diplomats, the astronauts and the presidents, for the one icon that did as much for the survival of a free world as any other: the American farmer.

Clearly, no political ideology, no economic system, can hope to be successful if it cannot feed its people. And there can be no doubt that one of the largest contrasts between the United States and the Soviet Union throughout their period of global rivalry was the success of American farming and the failure of Soviet agriculture.

That contrast certainly was one of the West's most-used propaganda weapons of the era. Every school child was familiar with the image of the robustly independent Jeffersonian yeoman farmer, exemplifying the virtues of the free market, against the drably automatonic Stalinist agricultural collective, drowning in bureaucracy and regimentation. The rugged U.S. individualist, with a mixture of freedom, profit motive, good old American know-how—and more than a little help from the government—

managed not only to feed his nation with the best and cheapest food on the planet but also to export enough to banish the specter of mass starvation from places such as India and Africa.

The Soviet farm worker, on the other hand, was little better off than a czarist serf, with no ability to make his own decisions about what or when to plant, no motivation to succeed, improve, or improvise to meet local conditions, and no ability to feed a nation. The Soviets may have had the first man in space. But America had its amber waves of grain and that, in the end, was more important.

So, as the Cold War recedes into history, what model of agriculture will the United States follow?

Marshal Stalin's model of centralized agriculture has not perished from the earth. If anything, it is slowly but surely taking over, in America and around the world. Except that now the label "collective farming" has given way to "vertical integration" or "production management." Even opponents of this trend—including me—often don't invoke a label any more threatening than "industrial farming" to describe it. George Washington, of course, was all in favor of farmers being industrious, but the kind of agriculture being practiced in America today is increasingly something he would not recognize. Nevertheless, old Kremlin hands would be quite familiar with it: centralized control of food production, decisions that no longer belong to those who work the land and feed the livestock, techniques that poison and exhaust the land, and a uniformity in the genes and breeds that bodes disaster when domesticated life stops evolving but the pests that threaten them do not.

Both doomed approaches, Soviet collectivism and American monopolism, are justified by the same one-word argument: efficiency. Food production, proponents of both philosophies posit, is far too important to be left to the unfocused thrashing about of millions of independent farmers who may strike off in all directions, only to come to the end of the season with grain or livestock that falls short of demand or, at least as bad, products that do not flow neatly through the processing and distribution chain like so many manufactured gears or lightbulbs. As if lecturing to so many loyal Communist Party members, corporate agriculture justifies its maddening passion to control, if not own, food production from seed to supermarket as a matter of quality control and industrial efficiency, supposedly ensuring the highest-quality product that can be processed and delivered at the lowest possible cost.

Lowest cost, that is, to the corporations that handle the various steps in the food-production system, before and after the activities that might still be somewhat recognizable as farming. This command-and-control system is certainly not low-cost when all factors are included in the equation. For example, it does not account for the environmental damage that forces everyone from rural well-owners to large cities to finance water filtration and purification systems—systems that would be unnecessary were it not for all the harmful nitrates that come from fertilizer and animal wastes. Nor does this system include the costs of health problems, and all the related medical care, lost wages, and human suffering, that result from the obesity, heart disease, and diabetes promoted by cheap, corn-sweetened, fatty foods and from the cancer and other diseases that are epidemic among farm workers who spend significant time handling chemicals. And it does not count the loss of tax base that states and communities have to make up, and the remote services they must provide, when rural communities stagnate and disappear for lack of independent farms to support and be supported by.

The number of people who decide how our food is produced, processed, and delivered decreases every day. The very DNA of the plants and animals we live on is being spliced, patented, and made the exclusive property of people we do not know and cannot vote for. The American farmer is being pushed, slowly but surely, into the role of his old Soviet counterpart. Even farmers who still own their land are pressed to use it in the service of factory chicken or pig producers, or for crops they must license from a faraway bureaucracy.

In a way that might make Uncle Joe Stalin smile, the current state of U.S. agricultural affairs might well be blamed on the old Soviet Union. As explained in 2002 by University of Missouri rural sociology professors William D. Heffernan and Mark K. Hendrickson at a seminar in Boston, with *The Onion*–sounding title "The Farm Crisis: How the Heck Did We Get Here," the decline and fall of independent agriculture in America could be said to have begun with what seemed like good news: the Russian Wheat Deal of 1972.

The sudden opening of that potentially huge market was only the most visible sign of what seemed a brave new world for American farmers. The increasing wealth of the nations of the Organization of Petroleum Exporting Countries (OPEC) also boosted world demand for American grain. American farmers, to their credit, were eager to meet that demand. Farmers resentful of the higher prices they, like everyone else, were paying for gasoline envied the power of the Arab oil sheiks, and many U.S. farmers proposed what sounded to many people, not only farmers, as a fair solution: a bushel of wheat for a barrel of oil.

Even if that had been a fair exchange (in the early 1970s, a barrel of oil went for $30, but a bushel of wheat rarely brought more than $4), it was not going to happen. The purpose of OPEC, then and now, is to boost the price of oil by controlling the supply.

Much of the world's oil comes not only from a handful of nations but also from nations in which the government controls drilling, pumping, and sale. Although oil-producing countries have a long history of sidestepping their self-imposed delivery quotas, important OPEC producers in the early 1970s were few enough and unified enough to limit oil supplies and boost the price. Meanwhile, in the United States, agricultural practice was precisely the opposite. A couple million farmers—too many, too diverse, and too cussedly independent to fix the price of anything—received no direction from the government other than to, in the words of Richard Nixon's later-infamous Secretary of Agriculture Earl Butz, "plant fence row to fence row."

The road from Stalin to Butz travels through *Animal Farm,* the classic George Orwell novel that wasn't really about animals but has since become far more about farming than its author probably could have realized. Orwell created a parable of the evils of Soviet-style socialism and how propaganda was used to confuse people about events around them and led them to think that anything wrong with their lives was their own fault, certainly not to be blamed on their leaders. The farm animal that most resembled the modern farmer—or the industrial worker, for that matter—was Boxer, the loyal, decent, and none-too-bright plow horse. No matter what woes befell the farm residents, Boxer was horse enough to shoulder his share of the blame and to resolve, "I will work harder." This sentiment won him the admiration of his fellow workers and, after he had worked as hard as any horse could, for much longer than any animal had a right to expect, a trip to the glue factory.

American farmers' version of "I will work harder," especially since the 1950s, has been "Get big or get out." Even farmers who had never heard of Boxer were aware that simply working harder, in the sense of manual labor and heavy lifting, would not get them anywhere. Instead, they tried to work smarter, which in the 1970s meant buying more, planting more, and harvesting more.

"I don't want to own all the land in my county," was the widespread crack recorded by farm journalist A. V. Krebs, "just that which borders my own property." He made the point that no farmer or any group of farmers wanted to corner the market in anything. Rather, it was the ever-present, and usually vain, hope that farmers could attain an economy of scale that would allow them to take maximum advantage of the market when prices were high and to survive periods of low prices by making up the difference in volume. Bankers, always American farmers' not-so-silent partners, of course wanted in on the bonanza, and they dangled easy low-interest loans in front of farmers to buy seed, fertilizer, or even more land.

Economists note that farmers often sink whatever money they earn or can borrow into more land and more equipment because the investment lowers their income-tax liability. That is particularly true when money is flowing and interest rates are low. But land and farm equipment are expensive, so even a farmer who has some positive cash flow likely has money only for a down payment and borrows the rest. In this practice, no money goes into the bank for a rainy day or, more likely in America, a drought-stricken summer. Instead, the farmer is left holding depreciating equipment and heavily mortgaged land, both of which are liable to be worth less than he owes on them because a depressed farm economy means there are fewer buyers.

University of Minnesota farm economist Richard A. Levins explains that farmers who are successful in the long term are not those who are good at growing crops or herding cows so much as those who are good, or lucky, at speculating in land. Those who buy land at too high a price, most likely with a heavy mortgage, are most likely to fail. Those who buy at a low price, or manage to pay off their mortgages and own the land free and clear, are much better able to weather the ups and downs of commodity prices.

The farm press rode the haze of the Russian Wheat Deal all the way up to 1980, predicting, Heffernan and Hendrickson said, "a glorious decade for U.S. agriculture." But the glory was short-lived—or never arrived. As early as 1977, the cost of producing crops was growing, and the price farmers received for their goods was shrinking. Higher prices for fuel, fertilizer, and machinery pushed the cost of producing a bushel of corn, for example, up to $3.55, whereas the market price, depressed by debt-subsidized overproduction, ranged from $2.06 to $2.80 a bushel. The problem wasn't that nobody wanted the corn; there were then, as now, many hungry people in the world. But global financial institutions, which had hoped that developing nations would industrialize and make money to buy American food, figured out that poor nations were still poor and stopped lending them so much money. American farmers stayed afloat by borrowing against their ever more valuable land, or seized the opportunity to cash out.

In late 1977 and early 1978, some farmers who figured out what was happening organized the American Agriculture Movement and mounted "tractorcades" to Washington to demand change. The TV cameras showed up to record images of John Deeres and Massy Fergusons surrounding the Capitol. It was good theater, but little changed.

By late 1979, all that cash flowing around the world had pushed inflation high enough that the U.S. Federal Reserve Board did what central banks tend to do in inflationary times: It raised interest rates—a lot, to levels approaching 20 percent annually. Farmers who had taken on heavy debt—the kind of debt that is refinanced every crop year—to buy more land and more equipment found themselves unable to keep up. Inflation had been the borrowing farmer's friend when interest rates were lower, because borrowing at low interest and repaying loans with inflated dollars amounted to a profit for the farmers. The reverse occurred as interest rates rose: The land purchased with those heavy loans was harder to sell in times of low commodity prices and high interest rates and thus was suddenly worth as little as half what was owed on it.

The number of farms and farmers had, of course, been declining for years, to the point that a nation once a land of farmers was now being fed by only 2 percent of its population. The percentage of Americans living on farms dropped from 32 percent in 1916 to 25 percent in 1930. After a brief uptick during the Depression, as many folks who could not find work in the city returned to the family homestead, the percentage of Americans engaged in agriculture continued to drop, to 23 percent in 1940, 15 percent in 1950, and a bit less than 9 percent in 1960. Academics and government officials who believed excess production could be resolved by reducing the number of producers never seemed to find that equilibrium.

The number of Americans engaged in food production has held more or less steady at 2 percent since 1985. The squeeze that still exists—higher costs and lower incomes—is expressed in increasing numbers of farm families either becoming part-time farmers, with day jobs in town, or finding creative alternatives such as organic farming or selling their beef, chicken, fruits, and vegetables direct to consumers.

The industrialization of America pulled people off the farms and toward the cities to work in factories and offices. The work and land they left were absorbed by those who remained behind and spent money—usually borrowed—on larger and higher-tech farm machinery, irrigation systems, pesticides, and fertilizer. In 1900 the U.S. Department of Agriculture (USDA) counted 5.7 million distinct farms in the United States, averaging about 147 acres each. A half century later, the number had fallen to 5.4 million and grown to an average size of 216 acres—a trifling difference compared with what was to come next.

By 1974 there were only 2.3 million farms registered on the USDA's Farm Census, and the average size had more than doubled, to 440 acres. In the five years ending in 1985, the counted number of farms dropped by 15 percent, from 2.5 million to less than 2.2 million. The government put the figure in 1997 at 1.9 million, but this was adjusted upward in early 2004 to 2.2 million to offset the number of farmers and ranchers who did not return the official survey. That same report posted the number of farms in 2002 at 2.1 million with average farm size up one acre to 441.

The long trend toward a smaller farm population differed from the severe crashes—the Depression of the 1930s and the farm crisis of the 1980s—in that the decline in farm population during the 1900s was relative to the enlargement of urban centers and was much more gradual. During much of this period, the farm population declined because farmers' children anticipated a better, more exciting life in the city so did not take over farms left by retiring or deceased parents. As the remaining farmers aged and died, they or their city-dwelling heirs sold or rented the fields, often to neighbors who enlarged existing operations. Banks and other financial advisers set themselves up to serve as managers of farmland owned by retired farmers, their widows, or their far-flung children and grandchildren. The demographic shift was less a matter of farmers going broke than of the simple evolution from rural to urban life.

Since the crash of the 1980s, though, the shift has been more painful and less voluntary. Farmers began going out of business, as they did in the Depression, due to low prices, high costs, and deep debt. Though Earl Butz was long gone by then, having committed political suicide by telling an off-color, racist joke, A. V. Krebs would not let him escape his share of responsibility for what happened to the American farmer: "Butz must now bear a major responsibility," Krebs wrote, "for the plight of thousands and thousands of American farmers who have been forced off their land and into bankruptcy, forced foreclosures, depression, divorce, alcoholism and suicide."

In the 1997 USDA Census of Agriculture (figures before the upward revision of total farmers), only a little more than half the 1.9 million counted farm operators claimed farming as their principal occupation. Of the 949,000 farms that were serious enough to make more than $10,000 in gross annual sales, 274,000 were run by operators who could not honestly list farming as their main means of support. Altogether, 709,000 of the 1.9 million farm operators worked 200 or more days somewhere else. Total farm net cash income that year was $55.7 billion. Total nonfarm earnings for farmers, mostly from nine-to-five jobs in town, were $124 billion. And increasing amounts of what was still called farm income came from contract growing of chickens, hogs, and cattle owned by or committed to the processing companies and simply raised by the farmers.

In the short run, it might seem as though avoiding the ups and downs of the global market by contracting to raise, at a preset price, somebody else's chicks or piglets from a little past infancy to a day or so before slaughter—with the processors selling the grower the approved feed and medicine—could be a good job. But, at least in America, most jobs generally come with some protections and benefits for the workers, such as a minimum wage, rules about working conditions, and health care coverage. Contract growers receive no such benefits. Further, jobs that require workers to invest thousands of dollars before they will even be considered for the work are usually the type of pitch that attracts the attention of a state attorney general's fraud division. Not so with contract growing.

There is no question that this practice started with the chicken, not the egg. In the new era of contract growing, the only eggs a chicken farmer may see are in the supermarket. He does see one or more truckloads of baby chicks sent from hatcheries pull up to his built-to-corporate-specifications barn, the one he went a half million or more dollars into debt to buy and equip. The load of chicks has been weighed so that when another truck comes to get the slaughter-weight broilers in about ninety days, the grower (this person could hardly be called a farmer) can be paid based on the weight the birds put on while they were in his care.

In his care but not his in name. Ninety-five percent of the chickens Americans eat are raised this way, by growers under contract to one of forty processors. Half those chickens—from egg to supermarket—belong to one of four giant food processors: Tyson, Gold Kist, Perdue Farms, or Pilgrim's Pride. They operate mostly in an arc of sixteen states from Pennsylvania south and west to Texas. In 2003 those states produced 7.5 billion chickens, 88 percent of the U.S. total.

Economists will tell you that when four firms control as little as 40 percent of a market (called "four-firm concentration"), those four—or even the dominant firm among the four—effectively have the power to set the price of a given commodity, without any true collusion or even contact among cartel members. They need only determine the price being paid by whichever other firm is buying animals in that region on that day—if any are—and match or only slightly exceed that price.

The poultry business structure today is the reverse of that in pre–World War II days, when most chickens were kept for the eggs they laid and were turned into dinner only on special occasions or when they were worn out as egg-layers (a situation mined to comic advantage in the animated feature film *Chicken Run*). There were 11,400 chicken hatcheries in the United States in 1934 incubating, at any one time, 276 million eggs. By 2001 there were 323 hatcheries keeping 862 million eggs warm at any given time. The shift to chickens bred specifically for eating— called "broilers" in the business no matter how they wind up being cooked—went hand-in-hand with the consolidation of the processors. In 1945 the United States turned out 1.1 billion pounds of chickens, compared to 19.5 billion pounds of cattle and 18.8 billion pounds of hogs. In 2001 broiler production soared to 42.5 billion pounds, more than the 42.4 billion pounds of cattle and 26 billion pounds of pigs.

Because the processing firms own the birds—as many as 90,000 of them in one barn—they understandably impose many rules on how a grower cares for their property. The chickens can be fed only a certain kind of feed and given approved medicines, purchased from the processor, and housed under proper conditions—all at the grower's expense. If disease or a heat wave decimates the flock, there is no pay for the grower. If the grower's neighbors complain about the odor, the state issues a citation for polluting the water, or the Poultry Liberation Front stages a dark-of-night chicken emancipation, all financial responsibility rests with the grower.

Growers sign contracts that often swear them to secrecy as to conditions and rate of payment. These confidentiality clauses mean growers cannot compare notes on whether they are being treated the same as other growers or are being paid the market rate for their contribution to the process. In fact, they don't even know the market rate. Should grower and processor become dissatisfied with each another, the processor can always find another grower, usually another farmer who figures that a promised rate per pound of added white meat beats the vagaries of the spot market. But the grower will probably find it impossible to switch to another processor, because there are so few of them and each has de-facto exclusive control over a given geographical area.

If the poultry processor decides, for any reason, that it wants to discontinue dealing with a grower, it simply doesn't send another load of baby chicks. Maybe the grower failed to live up to the contract or to some unwritten set of expectations. Maybe the standards set by the processor or by state environmental regulators have changed. Maybe the grower made a disloyal remark to the processor or, worse, about the processor to the press. In any event, the processor moves on. The grower is stuck with a heavily mortgaged chicken barn (or a row of such barns) and no way to fill it with chickens. Many former chicken growers have fallen into bankruptcy. Suicide has been the way out for more than one.

Because hogs and cattle are larger and take longer to grow to slaughter weight, the corporate takeover of those sectors (corporate reports refer to them as the "protein industry") has been slower to coalesce. Still, anything resembling a free market has been replaced by a system in which the four largest processors control the lion's share of U.S. pork and beef production. In the beef business, the cows are not the only things being slaughtered. The free market has been stunned and had its throat slit by a four-firm concentration of 84 percent. At the head of the herd is

IBP (once known as Iowa Beef Processors, absorbed by chicken giant Tyson in 2001) with a third of the market all to itself. The others at the top are Swift (spun off from ConAgra in 2002), Cargill's Excel subsidiary, and Farmland National. Smithfield is number five on that list, which is why a farmer cannot avoid the same agribusiness giants by switching from cows to sows. Smithfield slaughters more than 80,000 hogs every day, marketed under brand names that include Smithfield, Cudahy, and Ember Farms, to lead the list. Smithfield, Tyson/IBP, Swift, and Cargill/Excel among them control 64 percent of the national pork market.

The pork business is beginning to emulate the chicken racket in terms of livestock ownership. Smithfield alone owns more than 800,000 of its own breeding sows—14 percent of the nation's total population of the beasts. Along with the rest of the Porky Four, swine processors own 1.3 million breeder pigs. In both pork and beef, processors and farmers increasingly strike "forward contracts" that commit both sides to a specific price per pound of pig or cow on a specific date in the future. Like the poultry contracts, the prices paid to one grower are generally confidential, so nobody knows what "the market will bear." Transparency is impossible, because with no sale barn, no stockyard, no railhead, there is no market. Even if a Colorado cattle rancher could determine that a Texas buyer was paying top dollar for market-weight animals, that knowledge would do the rancher no good. The time and expense of shipping the cattle—including the weight the animals would lose en route—would eat up any profits. The processors, meanwhile, have such regional dominance, and so many animals lined up for delivery, that they are never against the wall of having to pay top dollar or search farther afield just to keep their production lines moving.

Vegetarians and grain farmers are not immune to this growing plague of monopolization. As with livestock, the growers of wheat, corn, soybeans, and rice find that the prices are set—low—by a handful of market-controlling buyers. Four firms—Cargill, Cenex Harvest Services, Archer Daniels Midland (ADM), and General Mills—among them own 60 percent of the nation's grain-handling facilities. Add Louis Dreyfus and ConAgra, and six outfits own 74 percent of the nation's grain bins.

Meanwhile, huge shares of the seeds that grain farmers buy at the beginning of the annual farming process come from one of four companies: Monsanto, Novartis, Dow Chemical, and DuPont have 75 percent of the nation's seed-corn sales and 60 percent of the soybean seed. Of course, the companies brag that their seed will outproduce the competition. For example, DuPont's Pioneer-brand seed is touted as "Technology That Yields"—despite the fact that overproduction is American farmers' worst enemy. Meanwhile, much of Monsanto's product line is genetically modified seed that, according to Monsanto, isn't worth having unless it is treated with Monsanto-patented pesticides. Thus, in keeping with the Sovietization of American agriculture, agribusiness giants are selling farmers the rope they are using to hang themselves.

By all these means and others, the ownership, genetic code, practices, and profits of agriculture are being collected in fewer and fewer hands—hands that have no dirt under the fingernails. If these profit-maximizing executives and their lab-coated minions never make a mistake, concentration makes no apparent difference to consumers in the supermarket, except that it keeps food coming at what seems to be low prices.

But if the new lords of the harvest ever guess wrong—as they are frighteningly likely to do in their quest to put short-term profit ahead of the long-term health of farmers, their fields, and all who consume those products—we will be teetering on the edge of a huge disaster, one that could very well lead to the famine and starvation that modern industrial agriculture is supposed to prevent.

FROM "MORE!" TO "TOO MUCH!"

> The system of agriculture . . . in use in this part of
> the United States is as unproductive to the practi-
> tioners as it is ruinous to the landholders. Yet it is
> pertinaciously adhered to. To forsake it . . . requires
> resolution.
>
> —George Washington

The romance of the proudly self-sufficient farmer is so much a part of the American ideal that the notion of a "federal farm policy" has always sounded downright oxymoronic to some ears. Many such ears were in the audience that enthusiastically greeted televangelist and sometime Republican politician Pat Robertson when he lectured at Kansas State University in Manhattan, Kansas, in October 1993.

In seeking simultaneously to praise his heartland hosts and belittle his liberal political rivals, Robertson extolled the brave free-market pioneers who had settled Kansas in the previous century without any help from a "Department of Covered Wagons in Washington." Cheers. Applause. And no inhospitable reminders (made public at the time, anyway) that Robertson was championing the free market on the campus of an institution that proudly claims (as does Michigan State University) to be the nation's first land-grant college.

Land-grant colleges—there are 105 of them now—grew out of the Morrill Land-Grant Act of 1862. They are a creation of the federal government designed to help the occupants of those covered wagons, once they put down roots, to grow successfully the food that would be needed by the developing industrial centers of the East. To promote successful agriculture, these colleges established research and outreach programs to improve plant varieties and to provide farmers with education and advice about breeding, irrigation, land erosion, and sundry other farming topics. Also in 1862, Congress passed and President Abraham Lincoln signed the Homestead Act, motivating westward expansion. The Department of Agriculture was created that same year. Federal action to encourage the development of the transcontinental railroad, designed to move immigrants west and the crops and livestock they raised back east, was also a major undertaking of that decade.

Incredibly, Lincoln and his Congress were able to pursue such massive new endeavors even as North and South were embroiled in the Civil War. But these steps toward a more active government support for agriculture were very much tied, coming and going, to that upheaval. The tension between regions over the issue of slavery was not based completely or even primarily on concern for the men, women, and children held in cruel bondage. Opposition to the South's "peculiar institution" of slavery, and particularly to its expansion into lands in the Southwest acquired in the Mexican-American War, was based largely on resentment held by farmers in the West and by mill workers in the North that the value of their toil was being unfairly undercut by the large pool of slave labor available to Southern plantation owners. But after the Confederate states pulled out of the Union, and their senators and representatives left Congress, the remaining delegates from the North and West were able to pass programs aimed at emboldening the small freeholder.

Although the moral dimension is not necessarily as obvious today, independent farmers are again handicapped by policies and practices that minimize the value of human labor and life, tipping the economic balance in favor of massive agricultural operations and away from human-scale farms that could prove their worth in a truly free, open—and fair—market. It is an equally peculiar institution, one that, in twenty-first-century jargon, might be called sharecropping 2.0.

While the slaveholding plantations, the industrial farms of their day, were busy with other concerns—such as how to fight an industrial war with no industrial base—the system set up by the Homestead Act offered a family 160 acres of federal land if they would live on it and farm it for at least five years. Alternatively, the government would sell them the land for $1.25 per acre, a bargain even by standards of the day, if they had lived on and farmed it for just six months. From 1862 to 1900, some 500,000 families used the Homestead Act to become landowners.

The Morrill Land-Grant Act, named for Representative Justin S. Morrill of Vermont, granted each state 30,000 acres of federal land. Each state was to sell the land and invest the proceeds toward the establishment and maintenance of colleges devoted to developing and distributing knowledge of the useful arts and sciences. That included such fields as home economics, manufacturing, mining, and, most prominent, agriculture. Some of those institutions began life under such names as Kansas State Agricultural College or the Agriculture and Mechanical College of Texas.

Today land-grant schools are often, but not always, recognizable by the middle name "State," as in Ohio State University or Oklahoma State University. In some states, such as Nebraska and Missouri, the entire state system of higher education is part of the land-grant system. So are some of the nation's more prestigious institutions, such as Purdue, Cornell, Rutgers, and MIT. Agriculture remains a core mission for many of these schools—they are headquarters for a state's system of county agents and soil conservation efforts and are the home of a few football teams still officially nicknamed "Aggies"—but the trend has been to drop any mention of agriculture from the institution's name. It's just not sexy enough. In the case of Texas A&M University, for example, the letters A and M officially do not stand for anything. They are merely legacies.

Homesteading and land-grant colleges were high points of long-standing federal efforts to increase the number and the output of American farmers, a federal farm policy that could be summed up in a single word, especially one followed by an exclamation point: "More!"

This trend dated to before the founding of the republic. In fact, the British Crown's unwillingness to encourage more immigrants to come to the North American colonies and to facilitate the efforts of those who did come to open new lands in the West—then defined as anything upriver of Philadelphia—was among the bill of particulars listed in the Declaration of Independence as among the actions "compelling us to this separation." Meanwhile, the fact that the uppity colonists wanted British soldiers to protect their rapidly expanding settlements from the Indians they were encroaching on, even as they violently opposed paying the taxes necessary to support those soldiers, was one reason the Crown felt so stung by the serpent's tooth of its ungrateful colonists.

Among the most ungrateful were George Washington and Thomas Jefferson, land-rich and cash-poor farmers, and John Adams and Benjamin Franklin, fat and gout-ridden as a consequence of living in a land where there was always plenty to eat.

In such ways, little has changed in the past 200-plus years.

Since the colonial era, American agriculture has been the story of figuring out ways to grow more food, not only by opening up new lands but also by getting more out of land already in production. At first it was nothing more than being smart enough to plant something that grew well in a particular soil; then borrowing Native American tricks such as planting a fish with a cornstalk to provide fertilizer; or planting different crops in rotation—corn one year, soybeans the next, grass for cattle to eat and poop on the third—so as to give the soil a chance to recharge. The twentieth century proved a dizzying march from no-tech to high-tech—industrial methods sometimes chillingly applied to the inherently biological process of producing food—including labor-saving machinery, fertilizers, pesticides, crossbred and, later, gene-spliced plants. Thus, the amount of food produced continued to grow to market-glutting proportions even as some of the land used to grow it was idled, retired, or just exhausted.

The westward migration and technical innovations that followed the Civil War were assisted by Homestead Act claims, land-grant colleges, federally subsidized railroad construction, and an Army—melded out of the battle-hardened surplus of Union and Confederate veterans—to ward off the Indians. Those increasing and increasingly productive farms fed the rapidly growing populations of industrial cities—not always well, perhaps, but enough to keep them standing through endless hours of factory toil. Some of those factories produced farming machines, such as Cyrus McCormick's reaper, patented in 1834, and John Deere's steel plow, invented in 1837. But even those marvels of nine-teenth-century technology had to be pulled by horses and mules, animals that ate the equivalent of one-fourth of the U.S. grass and grain crop. America's cities kept growing in relation to its farm population so that the 1920 Census for the first time recorded more city-dwellers than rural residents. All those people needed to be fed.

So did the hungry stomachs of Europe, especially during and after World War I. With French wheat fields turned into battle-fields, British farmers drafted into the army, and the flow of Russian commodities cut off from much of Europe, American farmers found themselves pushed to supply the largest and most far-flung market they had ever imagined. Soon the U.S. govern-ment's role in agriculture transcended its traditional boundaries of providing infrastructure—education, transportation, and se-curity—and took up the new exhortation to "Plow to the fence for national defense." For the first time, the federal government encouraged more production by setting minimum prices to be paid to farmers for basic food commodities, and it set up the U.S. Food Administration, headed by future president Herbert Hoover, to accomplish the job. Congress also added to the land-grant colleges a new system of county extension agents, whose job it was to deploy into the fields bringing farmers the latest agricultural innovations to help increase their yields.

The county agents are still out there, still helping farmers pro-duce more food. The idea that the federal government has a role to play in setting farm prices is still there, too, even if its efforts have often been, over the long haul, counterproductive at best.

After the 1919 armistice ending World War I, farmers enriched by wartime profits could buy more land, which they could work by shifting from horses to the new gasoline-powered tractors.

They also were able to afford more of the chemical fertilizers and higher-yielding hybrid seeds being developed by land-grant scientists and private industry. The farmers produced more and more grain—total production rose 13 percent between 1917 and 1929—and had fewer and fewer working animals to consume it. To be valuable, the grain thus had to find its way to people—city people—who could buy it. There were more of those, of course, but not as many as were needed to absorb all the food available. Europe also was regaining its ability to feed itself and no longer needed America to fill its larders.

The pattern of twentieth-century agriculture was set. Farmers who had money from a boom time invested it in land and equipment, often incurring heavy debt that, in a time of high income, seemed perfectly reasonable to farmer and banker alike. The national and world economic growth necessary to provide a paying market for all that food continued, but not as fast as farm production. Depression—the existence of way too much stuff to buy and not enough money to buy it—was the result, striking the farm sector of the economy well before the stock market crash of 1929. Farm population continued to decline, both in raw numbers—from 32.5 million to 30.5 million between 1916 and 1930—and as a percentage of the total population—from 32 percent to 25 percent over the same period. Those who remained on the farm were squeezed by the need to buy retail and sell wholesale.

As early as 1922, Secretary of Agriculture Henry C. Wallace pointed out the folly of ever again trusting farming to the tender mercies of the free market: "It will never be possible for the farmers to relate their production to profitable demand with the nicety of the manufacturer, both because they cannot control the elements which influence production and cannot estimate demand as closely."

The situation touched off more serious political wrangling over the appropriate federal role in farming. By 1929, Herbert Hoover, whose part in boosting U.S. farm production was credited with averting widespread famine in Europe during and after World War I, had become president of the United States. His "Food will win the war!" experience, coupled with the innate human belief that there can never be too much food, moved him to hew to a market-based—or, more precisely, marketing-based—solution. Hoover's 1929 Agricultural Marketing Act created the Federal Farm Board, which was to make loans out of a $500 million revolving fund to newly formed farmer co-ops. Members of those co-ops were supposed to work together to withhold their grain from the market when prices were low and find buyers when prices were high.

Created on the very eve of the Great Depression, Hoover's farm marketing solution never had a chance. Widespread joblessness and poverty caused such a drop in demand for all goods that gross U.S. farm income fell 52 percent between 1929 and 1932. Grain exports evaporated, and the price of wheat and corn, the underpinning of the nation's agricultural economy, plummeted. Wheat plunged from $1.03 a bushel to 38 cents. Corn fell from 80 cents a bushel to 38 cents.

Faced with such a hopeless market, a manufacturer of cars or radios or shoes would cut back on production, both to save on expenses and hope that a reduced supply of anything will eventually balance with the reduced demand to boost prices. That works particularly well, for the business if not the consumer, when an industry is dominated by a few key players—Ford, GM, and Chrysler, say, or Microsoft and, well, nobody. Compared to farmers trapped by biological cycles, other big businesses can ratchet down production with relative ease whenever the market wanes, then crank it up again when demand for its old product increases or it finds a new product with which to tempt the masses.

Farmers have never been able to do that.

It's not as if the flow of wheat or meat from the farm were like diamonds in the mine or oil at the wellhead, controlled by a few powerful people who can turn the supply on and off as the market demands, or simply as it suits them. Even the makers of more pedestrian goods can respond to the feedback of the marketplace much more quickly than can most farmers. Declining sales of anything—from cars to toothbrushes—can be and often are answered by a deliberate decline in production, or at least in distribution. The idea can be not only to cut costs but also to juice up demand for the product by pushing people to think it has become rare and might become unattainable.

The process of bringing a crop from seed to harvest cannot be abandoned so easily. A field of wheat cannot start producing sunflowers in midseason, even if sunflowers are what the market wants to buy. Excess farm inventory cannot be cleared by cutting prices, as prices are generally below the cost of production already and are set by national and international markets, not by anybody's marketing department. Even if a farmer had the guts or the financial reserves to respond to a period of low prices by skipping a season's planting—or could afford to buy or rent enough personal storage capacity to hold a crop off the market until the price went up—he would have to be joined by many thousands of other farmers in order to make much difference in prices that are set by national, if not worldwide, markets.

Cutting back on acres planted, or just farming them less aggressively, would mean less expense for seed, fertilizer, pesticides, and such. But it would not make a dent in the much greater fixed costs a farm carries, mostly the mortgage on the land and the payments on the tractors, combines, trucks, and irrigation equipment. Those things cost money whether they are put to full use or not, so the natural tendency is to get the most that one can out of them. Another problem is that most independent farmers, unlike factories or even large retail operations, do not have the option of cutting costs by reducing the number of workers. Most farmers have little or no workforce to lay off. Farmers do the work, along with members of their families. Even larger farms tend to be run by few people operating bigger machines. When no labor costs appear on a balance sheet, no cutbacks are possible there to save money.

When commodity prices are high, farmers want to produce as much as possible to reap maximum profit for each bushel or ton they produce. When commodity prices are low—which they will eventually be when every farmer produces as much as possible— farmers are still moved to produce to the limit of their capacity because when the per-bushel price is down, they need to sell more bushels to cover the fixed costs. So production stays high, and as a direct result, the value of that production stays low or goes lower. That was the setting in 1931, when Depression America produced the third-largest wheat crop in U.S. history up to that time, an accomplishment that was rewarded in 1932 with a market that paid a rock-bottom 20 cents a bushel.

The 1932 election of Franklin D. Roosevelt and the creation of his New Deal were turning points in any number of policy matters, perhaps most significant being the way government addressed food production. Washington thereafter recognized, at least in theory, that the sign on the wall could no longer be Hoover's— and Lincoln's—"More!" It became Roosevelt's "Too much!"

One highlight of Roosevelt's whirlwind first 100 days was the Agricultural Adjustment Act of 1933. Today it would be called a major attitude adjustment, or a paradigm shift, as it recognized that the woes of the farm economy were caused almost wholly by its collective habit of overproduction coupled with the individual farmer's inability to cut back on production and still have any income at all. The problem was not that we were running out of food. It was the threat that we might run out of farmers, and then we would run out of food or, more likely, see the role of food production taken over by a few corporations or foreign producers who would hold American consumers up for prices that were unaffordable even in good economic times.

The new Agricultural Adjustment Administration existed to do the previously unthinkable. It paid farmers to produce less food. More accurately, it gave farmers money so that they could afford to produce less food. The idea was that if farm families could stay afloat, even for a little while, on government payments, they would not be spurred to max out production just to keep from losing their farm to foreclosure. In practical terms, the new policy meant leaving a certain percentage of a farmer's land unplanted in return for federal payments or the right to participate in some other federal support program. (In the emergency situation of 1933, it went beyond that to the destruction of 10.4 million acres of already-planted cotton and the slaughter of 6 million hogs that were made into lard, fertilizer, and some food donated to the poor.)

Do that with enough farmers, the theory goes, and the supply will eventually shrink to match demand. That, in turn, will push prices up enough that farmers can better live off what they produce and their dependence on Uncle Sugar to keep the wolf from the door will eventually disappear or, at least, shrink to levels that are not a significant burden to taxpayers. But the country has followed this line now for seventy-six years. That day hasn't come, and without another serious shift in thinking, it won't come. Farmers cannot escape this dependency because twentieth-century farm policy, not yet significantly changed for the twenty-first century, remains one that encourages American farmers to produce more than the market can absorb. That keeps prices low and perpetuates the need for government assistance. The problem now is that, unlike during the spell from Lincoln to Hoover, pushing production to its maximum is an unintended and damaging consequence.

Since 1933 federal farm policy has used various methods, programs, agencies, and formulas to move toward its overall goal. That goal is not to increase food production but to reduce the supply of the so-called program crops—primarily wheat, corn, soybeans, rice, and cotton—enough to raise the market price to a level that will allow farmers to make a decent living. But there is a corollary: not to raise the price so high as to make food unaffordable for the rest of us.

It made sense to focus on those crops when subsidy programs began during the Depression. Basic grains and cotton were in surplus, then and now, because they could be commonly and easily grown and, once grown, easily stored and shipped long distances. A surplus of those crops was most likely to remain a surplus, with local shortages easily overcome by great quantities brought from elsewhere. Fruits and vegetables, on the other hand, were perishable and therefore not as easily stockpiled or shipped, at least in the days before widespread refrigerated containers hauled them long distances in short order via improved highway and air transportation. But government habits, once formed, can be extremely hard to break because they breed established lobbies that fight hard to maintain habits.

The two basic methods of subsidizing program crops have

been direct federal payments to farmers and federal loans. Direct payments have been calculated in different ways that usually have something to do with the price the grain is fetching on the open market, the price the government figures it should fetch, and how much it cost to produce; then a government payment makes up the difference.

Government preplanting loans have also been made on the assumption that once the harvest is in some months hence, it will command a particular market price. If the market price falls short of the target price, the farmer forfeits his crop, which was the collateral on the deal, to the government, and does not have to pay off the loan. That supposedly leaves the farmer with enough cash to keep the farm and try again next year. The grain the government gets that way, along with surpluses it sometimes buys outright, can be stored or fed to the poor here or anywhere in the world. Either keeping or giving, at least in theory, has the effect of taking the commodities off the open market, where they were never so likely to have become life-giving meals as to have been price-depressing surpluses.

Whether the method of support is payments or loans, a condition for participation traditionally has been for farmers to limit production and/or take a certain share of their land off-line for at least one growing season and perhaps much longer. This arrangement also is supposed to reduce the amount of grain and cotton available, so buyers will pay more, taxpayers will pay less, and farmers will come closer to standing on their own.

The theory has never quite worked out in practice, however. The incentive to reduce production is incomplete, and almost all innovations out of government and university research, private-sector labs, and tinkerings of innumerable serious farmers and hobbyists alike are aimed at increasing the amount of food produced. More productive seed, whether crossbred in the old-fashioned way or bioengineered, has been treated with more fertilizer, inoculated with more bug-killer and weed-killer, and bathed in more water pumped from deeper in the earth, drawing more food out of less land and netting farmers a lower price for each bushel produced.

Meanwhile, most land that farmers have set aside—for a season, a year, or a generation—has been land that wasn't very productive in the first place. Giving up the effort to wring more crops out of such soil was smart for the farmer, who made out better getting paid to ignore land that had been sucking up extra labor, seed, and fertilizer just to produce marginal amounts of food. It also was smart for the federal goal of fighting soil erosion, as the marginal farmland being retired was often the land most likely to be washed away in the next hard rain, taking tons of chemicals into the water supply with it, especially if it had been laid bare by plowing and the planting of annual crops that never held the soil in place the way a permanent stand of grass would. But idling some land also meant that farmers diverted more of their effort and resources into already highly fertile land, making it even more so and, again, depressing prices.

USDA records that date to 1866 show that the peak years for acres devoted to corn ran roughly from 1909 to 1918, with more than 100 million acres harvested in the United States annually. With a yearly average yield of 22–29 bushels per acre, the nation's output of corn ran 2.5–2.9 billion bushels annually. Even before the demand associated with World War I drove up the price, corn in those years fetched an average of what was then considered the princely sum of 59 cents a bushel. Over the years, land devoted to corn decreased due to diversification, federal support, and urbanization, dipping as low as 54.5 million acres in 1969. But yields by that year had climbed to nearly 90 bushes an acre, and total production topped 4.5 billion bushels. By 2003 farmers perceived that corn was once again where the secure money was, and harvested acres were back up to 71 million. Yield soared to 142 bushels an acre, putting 10 billion-plus bushels in the bin. Corn prices in 2003 were $2.45 a bushel.

That looks a lot better than 59 cents, right? Not so. According to the Agriculture Department, the 2004 price of corn would have to have been $6.55 a bushel for it to have kept pace with the average farmer's pre–World War I buying power.

Those calculations are a symptom of the fact that throughout the twentieth century, U.S. farm production increased faster than the number of Americans available to eat it. In the 1930s, production was up 12 percent, but population was up only 7 percent. In the 1950s, production averaged a 2.1 percent annual increase; population rose only 1.7 percent a year. And production increased even though fewer people were raising the food.

In 1996, Congress set out to do something completely different. Formally known as the Federal Agricultural Improvement Act (FAIR), the legislation was known to its friends, especially chief author Senator Pat Roberts (R-Kansas), as Freedom to Farm. In concept, Freedom to Farm took a clear-eyed look at what was wrong with every farm bill since the New Deal and tried to change it for the better.

Because direct payment and loan programs over the previous sixty-three years had been tied to specific products—usually wheat, corn, rice, soybeans, and cotton—farmers either planted those crops or didn't get to play. Thus those crops were grown in excess, year after year. That made (and makes) sense from the point of view of both consumers and middlemen—the latter including the feeders and processors of the animals that eat the grain before people eat the animals. Wheat and rice feed people the world over, corn and soybeans feed cattle first and people second, and cotton covers billions of bodies. The premise of keeping those products plentiful and thus cheap was logical for the families, cattlemen, bakers, chicken growers, grocers, butchers, and the makers of Corn Flakes, Rice Krispies, and Wheaties.

Logical, however, only because these food middlemen never stopped to consider that what farmers had long derided as Washington's "cheap food policy" actually had some serious costs to it. Farmers were losing their denim shirts. Taxpayers were spending billions to keep farmers from losing their overalls, too. And continually encouraging maximum production, even if that was not the stated goal, encouraged maximum use of expensive and dangerous chemicals and practices that were exhausting the soil's long-term health for short-term gains.

Freedom to Farm's approach was to "decouple" a farmer's choice of crops from federal program rules. Instead of their participation being tied to specific crops, farmers who had relied on federal payment and loan programs in the past could receive seven years of declining transition payments and plant whatever crop they thought the market would want, in whatever amounts they thought the market would bear. Although there were provisions to pay farmers for returning to its natural state some land in areas particularly susceptible to erosion, there were no setaside requirements or other production limits for those receiving the basic payments. The reduction of land for crops had never significantly reduced the overall amount of grain produced, Senator Roberts noted, so why bother?

Freedom to Farm became law in 1996. By 1998 it was being judged a complete failure.

Planting the basic grains was no longer a requirement of federal aid. But they were still the crops the worldwide food-processing system was set up to handle, farmers knew how to grow, and agribusiness knew how to buy and sell. There was no marketing blitz for Barley Flakes, no SpongeBob SquarePants–brand lentils. Cows still ate corn and soybeans, Americans in Asian restaurants as well as Asians in Asia still ate rice, and everybody still ate bread made from wheat. So corn, soybeans, rice, and wheat were still what American farmers grew. And, with no requirement that farmers limit their production or leave part of their land as pasture, they all grew more and counted on the already guaranteed federal payments—$20.5 billion in 1999—to cover their costs.

A type of Asian flu undid the deal, though. Not the fever-and-chills flu, but the sharp economic downturn that dulled expected Pacific Rim demand for American farm products. The Asian market had been a big buyer of U.S. farm products in the years before Freedom to Farm was passed, and its initial popularity among farmers and policymakers alike was based on the assumption that Asian demand would go on forever. Instead, exports declined even as U.S. production increased.

In fact, not only have exports of U.S. farm products never been sufficient to deplete the nation's surplus, except in the occasional bonanza year that blinds everyone to reality, but imports of farm goods into the United States are growing steadily as well. These include both nonnative items—bananas, coffee, rubber—and even wheat, corn, rice, and other imports that compete head-to-head with U.S.-produced goods. In 1990, according to USDA figures, the United States exported $39.5 billion in agricultural goods and imported $22.9 billion worth. Since then, exports fluctuated up and down year to year, but imports moved only one direction—up. By 2002 food exports were $53 billion, but imports had closed the gap to total nearly $42 billion, only $7 billion of which were so-called noncompetitive goods not grown domestically.

Freedom to Farm's declining transition payments that were supposed to tide farmers over to a bright new day of free enterprise were suddenly not enough to keep a farm operation going when the price of what those farmers were selling was slumping—again. Congress rode to the rescue—again—with a new round of subsidy payments calculated under the rules of the 1938 version of the AAA, which had been overlaid many times but never repealed. These "emergency payments" were, as in old times, based on the amount of crops a given farmer had produced in the recent past. The larger producers, as usual, got the most money, reminding everyone involved that even when the government says it won't reward overproduction, it finds a way to reward overproduction. By 2001 Congress had provided farmers $71 billion in phase-out and emergency payments—three times the projected budget for Freedom to Farm.

When it was time to renew or replace FAIR—farm bills traditionally expire in five or six years—there was little fight left in the idea of having a real free market in American agriculture. The 2002 Farm Security and Rural Investment Act was an unabashed return to the New Deal, promising farmers $82 billion in added subsidies over ten years so that their eternal excess production would not drown them. The law also included financial incentives to retire some land that would otherwise be used to produce grains, both to limit production and protect sensitive lands. President Bush and others expressed continued belief in the ever-elusive promise of soaking up U.S. food surpluses by aggressively marketing them abroad, even though the hopelessness of that proved to be the Achilles' heel of Freedom to Farm. At base, though, the 2002 law was another congressional deferral to the notion that government's role should not be to end the U.S. farm sector's constant state of "too much!" but merely to manage it. A tough enough job, arguably, but certainly an expensive one.

Senator Roberts went on to become chairman of the Senate Intelligence Committee, where he could contemplate the comparatively simple problem of how to end worldwide terrorism. His lamentations about the fate of his Freedom to Farm brainchild, meanwhile, deserve to be placed on the record. Roberts never claimed or expected that simply releasing the pent-up energy of American farmers, sending them into the great beyond of the world marketplace, would solve all the problems of food production. To him, Freedom to Farm never really failed because it had never really been tried. Parts of the theory were never implemented, but he insisted they should have been if the proposal were to be given a fair test, much less have a chance to succeed.

One Freedom to Farm idea that should have been pursued, and still could be, was to take steps to soften the boom-and-bust nature of farming for individual practitioners. One solution would be a sound system of income insurance that would be producer-funded and would pay out to farmers who had no crop, due to bad weather, pests, or other misfortune, or whose crop had no value due to a collapse in farm prices. But that suggestion has never gone anywhere because farmers have a half century of experience telling them that the two disasters they fear—collapsed prices due to oversupply or ruined crops due to bad weather—are generally covered by Congress, which votes another round of emergency relief for farmers in such a pickle without requiring that any premiums be paid by the farmers who benefit. Such payments, all from taxpayers' money, ran to $2.4 billion in 2001 alone and totaled $11.3 billion in the years 1995 to 2003.

Another solution would be to create a tax-sheltered savings

plan to accommodate the cyclical nature of farm income. It would be a sort of 401(k) for farmers, accounts in which they could shelter profits made in good years and draw on these funds, without tax or penalty, in lean times. Such a program would be a superior alternative to the trap American farmers too often fall into when deciding what to do with any windfall, the trap of investing it all in more land or more heavy equipment. Indeed, adopting the same kind of matching contribution offered by many employers for their employees' 401(k) plans, paid by the government up to a certain percentage, would almost certainly be a better investment for the taxpayers than the current system of subsidies.

The money tied up in land and equipment not only is a key incentive for farmers to produce as much as possible and live with the deflating effect that overproduction has on commodity prices, but it also is a key to the willingness of Congress and the agriculture lobby to continue favoring the very subsidy programs that manage rather than reduce overproduction. If federal subsidies ever disappeared completely, a huge chunk of farmers' income would go with them. If farming no longer paid, farmland would suddenly be worth a lot less money. If farmland was suddenly worth a lot less money, farmers would lose their investment in that land, which is often the only nest egg they have to pay for their retirement and/or to leave to their children.

Other parts of Roberts's Freedom to Farm project were just as well consigned to the dustbin. They included reduced governmental regulation; increased research and development into ways to make farming more efficient; and renewed federal efforts, at the highest level, to boost the amount of American farm products sold overseas. All would be aimed at easing the squeeze on farmers caused by higher production cost and lower selling prices. None of these measures was likely to make much positive difference.

Decreasing production costs by allowing farming to become dirtier is unlikely to bring any savings to consumers, because cost of the actual farm product is a minuscule part of the cost of packaged food. In fact, such action would increase the hidden cost of cheap food by causing more pollution for others to endure or, at least, pay to clean up. Scientific research could be beneficial, if that research were geared to making farming cleaner and its products healthier, rather than adhering to the current corporate mind-set of pursuing ever-greater production.

Selling more American farm products overseas is a noble goal. Feeding the world, after all, is a worthy cause. But increased exports are unlikely, as most nations either have their own food or lack enough money to buy from America.

BUT WHO WILL FEED
THE WORLD?

In reality, only for today's affluent First World citizens, who don't actually do the work of raising food themselves, does food production (by remote agribusiness) mean less physical work, more comfort, freedom from starvation, and a longer expected lifetime. Most peasant farmers and herders, who constitute the great majority of the world's actual food producers, aren't necessarily better off than hunter-gatherers.

—Jared Diamond, *Guns, Germs, and Steel:*
The Fates of Human Societies

"Don't talk with your mouth full."

That is the likely response if you dare to question the way the vast majority of the world's food is produced today. It doesn't matter if you are worried about the nutritional value of the modern American diet, sickened by the images of feedlot-raised and factory-processed meat, doubt the wisdom of poisoning soil and water with every chemical science can devise, or frightened by the undiscovered country of genetic engineering. The response to each of those concerns, and a great many others, is basically the same, whether the person responding is wearing bib overalls, a lab coat, or a three-piece suit:

"We're the ones filling the supermarkets with food you can afford. Don't ask questions."

The concept is expressed in many ways. It might be the farmer's heartfelt claim that any change in practice or policy will push him over the edge of bankruptcy, an edge upon which his family has been teetering for generations. It might be the research scientist's assurances that any risks associated with the latest examples of human cleverness are minimal, or at least are outweighed by the promise of increased production to feed a hungry world. It could be the salesmanship of the corporate executives, and the legislators and regulators who are in their thrall, who habitually remind us that Americans enjoy the safest and cheapest food supply the world has ever known.

The it's-our-way-or-worldwide-starvation argument presents a false choice. Nevertheless, as intelligent and concerned people surveyed the world in the twentieth century, they saw reason to worry that mass starvation was a very real threat. From the early 1960s to the late 1990s, the population of the world nearly doubled—from about 3 billion people to 5.7 billion. The earth did not grow in size, and the portion of it devoted to growing crops increased only 12 percent. Immense population growth, countered by only slight growth in cultivated land, led number crunchers to calculate a net decrease of 42 percent in the amount of cropland available to feed each person.

So, why didn't we all starve to death?

The Green Revolution.

Developments in plant breeding, fertilizers, pesticides, and irrigation produced a boom in food production such as the world had never seen—necessary if we were all to avoid the Armageddon-like predictions included in books such as *The Population Bomb* and *Famine, 1975!* In the thirty-four years between 1963 and 1997, the average amount of crops produced per unit of cultivated land doubled worldwide.

This net gain in the planet's ability to feed humankind was reflected in the declining numbers of people living in hunger, particularly in the developing world. At the beginning of the 1970s, some 960 million people, or 37 percent of the developing world's population, were undernourished. By 1997 that figure had dropped to 790 million. That's still about 790 million too many, of course, and the number began rising again, topping 800 million in 2003. Still, the total number of Third World hungry declined, even though that same portion of the world was home to most of the planet's population growth. The percentage of Third World residents who were undernourished dropped from 37 percent in 1971 to 18 percent in 1997. The United Nations estimates that the absolute number of hungry in the developing world could fall below 600 million—or 10 percent of those nations' population—by 2015 and to 400 million, or a mere 6 percent of the population, by 2030.

Between 1962 and 1996, world crop production was up 117 percent. In every category, the Third World as a whole found a way to produce more of its own food. Total production in the developing world was up 175 percent during that same period. The amount of grain available for export from the three "developing exporters"—Argentina, Thailand, and Vietnam—rose from 12 million tons in 1975 to 23 million tons in 1996. The United Nations Food and Agriculture Organization (FAO) expects that number to increase to 32 million tons by 2015. The so-called transition countries, the various nations of the old Soviet bloc, are rapidly moving from being net importers of grain to being net exporters. Meat production worldwide also skyrocketed, from 81 million tons in 1965 to 151 million tons in 1985 and to 203 million tons in 1996.

All this output must be counted as a victory, to be sure. Anything is better than mass starvation, plus all the migration, social upheaval, epidemics, and wars that would certainly accompany any significant increase in the number of hungry people. Thus, when the scientist whose work was credited for sparking the Green Revolution, American agronomist Norman Borlaug, was awarded a Nobel Prize in 1970, it was not one of the science prizes. He received the Nobel Peace Prize.

Still, this substantial victory has come at substantial costs, costs that have been driven by political and economic factors at least as much as by any changes in crop science. In fact, Borlaug could be rightfully compared to better-known atomic geniuses such as Albert Einstein, Leo Szilard, and Enrico Fermi, whose understanding and development of nuclear energy were co-opted to become the property of governments and industries that would happily sell a hundred Nobel Peace Prizes for short-term gain.

Even accepting the argument that the use of the atomic bomb on Japan was ethically acceptable because of the greater harm it prevented—a cataclysmic invasion of the Japanese homeland—it is clear that the aftermath of that decision created a great many problems that have yet to be resolved, from sloppily run nuclear power plants to the possible use of nuclear weapons by terrorists. The same is true of the Green Revolution, an atomic bomb of sorts that prevented mass starvation in the near term but created a long list of problems for the future.

The excesses of the Green Revolution—all revolutions have their excesses—flow from the fact that the new generations of plants it created flourished by being better able to turn the energy from fertilizer and water into edible seeds—mostly wheat and rice—rather than into mostly useless stalks or hulls. But in many cases, the miracles of the revolution would not have been attainable except that farmers in the Americas, Europe, and India began to irrigate more. That usually involves buying pumps and the fuel to run them, as well as drawing down precious underground water tables to bring up water that rapidly evaporates in arid climates, doing little for the crops and leaving fields covered in leftover salt.

This ineffective irrigation meant farmers had to apply more expensive fertilizer, which is energy-intensive to manufacture and transport and has many environmental downsides. During roughly the same years that total world crop production rose 117 percent—the mid-1960s to mid-1990s—fertilizer use worldwide soared almost fourfold, from 34 million tons to 134 million tons.

The increased production made possible by these practices has only served to depress the prices farmers receive for what they produce, while increasing the cost of growing that crop. Huge sums of taxpayer money go to subsidize American and European farmers, while governments and private charities still struggle to feed people in regions where local food production and marketing have broken down due to natural disaster, war or, more often, a flood of cheap imported food.

In poorer countries, farmers may receive no assistance from their impoverished or corrupt governments. Third World farm families, already squeaking by on less than a dollar per person per day, cannot afford irrigation, fertilizer, or heavy machinery necessary to take full advantage of the Green Revolution's bounty. The worldwide market price on the crops those bystanders can produce is depressed by the surpluses generated elsewhere, so farmers who lack both government subsidies and the means to buy Green Revolution tools are left behind.

At the same time that food production was increasing, human population growth was decreasing. There are not fewer people—only a worldwide cataclysm would cause that—but the rate of population increase has slowed considerably. In 1993 the FAO's best estimate of how many mouths there would be to feed in 2010 was 7.2 billion. By 2000 FAO had revised the estimate downward by 400 million people, to 6.8 billion. Fewer people, of course, should mean more food to go around or, more important, potentially more money to go around, money that will make it possible for more people to buy food and perhaps even enable farmers to make a living growing it. Of course, that scenario presumes the distribution of money among would-be buyers is equitable enough to allow it to flow to more farmers.

Why will hungry people be fewer than the experts once predicted? There are two primary reasons. One of them is a historic change in human culture that, with luck, will continue to thrive. The other reason is monumentally dreadful, one that threatens to worsen in the future and destroy all hope of a well-fed Third World.

The good reason for the decline in population growth has been the relative shift in fertility—toward plants and away from human beings—and a shift in power—toward women. This is a cycle that builds on itself and, even with the risks that increased agricultural production can carry, can work to the betterment of all. It is a highly Darwinian feedback loop that works like this:

When there is too little food, or even when there is barely enough, rural cultures see great value in women having as many children as they can. The children are necessary to maintain the labor-intensive activities of hunting and gathering, slashing and burning, sowing and reaping. There is uncertainty about tomorrow's meals, and there is uncertainty that any of the children will live long enough to be productive, much less long enough to support their parents in their old age. People thus choose to have and, as best they can, maintain large families with many mouths to feed, even when infant mortality is high, all teetering on the brink of starvation. So there is never quite enough food. People then have more and more children, ensuring that there will still not be enough food. And so on. And so on. And so on.

But when there is enough food, and when there is confidence that there will continue to be enough food, that deadly cycle can be broken. Food supplies that had to be stretched to feed a family of eight go quite nicely around a family of four. The children, even the girls, have a chance to go to school. The young women have the opportunity to engage in the single most important step in human evolution: They marry for love.

Instead of being sold to another family as breeding stock, a young woman in a food-secure culture can be selective in her choice of a mate, even leave the bastard if he doesn't measure up. Thus more men are required to stay home and be good providers, to contribute to the family unit. These pressures push cultural evolution higher up the ladder to civilized and sustainable living that can support, and be supported by, a less industrialized form of agriculture.

The dreadful reason for slower population growth is AIDS. The disease is threatening India, Russia, and other developing and re-developing nations and, at the beginning of the twenty-first century, was hitting hardest in sub-Saharan Africa, a part of the world that would otherwise have been the prime engine of world population growth. The U.N. estimates that the total population of twenty-nine nations in southern Africa will reach 698 million by 2015, or 61 million people fewer than there would have been without AIDS. In the short term, this plague reduced the number of hungry mouths to feed, but in the long term, and for the foreseeable future, it has reduced the number of farmers who will be able to feed them. The Green Revolution allowed more people to be fed by fewer farmers, or more urban families to be fed by fewer rural families. But AIDS has deprived those rural families of their primary pool of both knowledge and labor. It has built a growing population of aged parents, widows, and orphans who, even if they do not lose the land their husbands and fathers and grown sons worked, lack the strength or the knowledge to farm it themselves. Unlike childhood diseases that previously, and cruelly, kept population growth in check, AIDS does not cull the number of small children to be fed so much as the number of able-bodied adults able to feed them.

AIDS also serves as an example of a spiral of problems that can build when long-standing, even economically fragile equilibrium is disrupted. Men who no longer can support their family by farming often flee to the cities in search of work. There they encounter women who, also due to the poverty that comes of a collapsed rural economy, live by selling their bodies. Especially in cultures where women still have little say in sexual matters and are unable to resist sexual advances or to insist on the use of condoms by their partners, the spread of disease is assured.

The support for dead-end, land-destroying practices often comes from otherwise responsible and environmentally concerned farmers because they have been beaten down by the big lie, the lie that hitching their wagons to the star of multinational food manufacturers will allow them not only to survive in this globalized market but also to continue their holy work of feeding the teeming masses worldwide.

In truth, such thinking amounts to American farmers slashing their own throats and those of the Third World unfortunates they take such pride in feeding. Given the high suicide rate of farmers from Iowa to India, that is not just a figure of speech.

The well-intended and often Herculean efforts to respond to severe famines, such as the 1985 Live Aid concert organized to meet an African catastrophe, doubtless save thousands of lives and remind the developed world of its responsibility to the rest of humanity. But after the glory fades and the news cameras focus elsewhere, the underlying causes of such disastrous food shortages are not remembered, much less addressed. Those causes are very much tied up in the overproduction mind-set that bedevils American agriculture.

Big agriculture is wedded not only to the idea that the world can grow its way out of its problems, but also to the idea that most of that growth should be in America—in the American boardrooms, anyway, if not on American farms. Archer Daniels Midland (ADM), one of the surviving handful of globe-girdling food merchants, used to call itself "Supermarket to the World" and continues to sketch its mission as helping to create America's agricultural bounty and then selling it to people around the world who otherwise would be very hungry indeed. The global reach of the corporation allows it reasonably to call its website "ADM World," on which it pledges to "advance the battle to end global hunger" by, of course, selling its products worldwide. The slogan of the Kansas Farm Bureau, "Helping to Feed the World," leavens its multinational arrogance with typical Prairie modesty: We're only helping, don't you see. But the KFB logo—seen on traditional seed caps and modern casual-day polo shirts—recalls nothing less than the United Nations emblem, with the olive branches that flank the globe replaced with stalks of Kansas wheat that encircle the earth at its waist.

Government subsidies, debt, and off-farm income allow American farmers to stay in business even while selling their grain for less than it cost to produce. The handful of companies that buy the grain from the farmers and resell it around the world have such a corner on the market that they never have to bid up the price of corn, rice, or cotton to wrest it from any competitors. ADM is one of three corporations—Cargill and Zen Noh are the other two—that control 80 percent of U.S. corn exports. So U.S. exporters sell cheap grain and cotton around the world. They make the profit and the farmers and the taxpayers who subsidized them feel good about feeding the Third World.

But the sad truth is that the United States, and the other food-flush nations of the world, do the hungry no favors by seeking to flood the planet with foodstuffs, animal feed, and cotton. If anything, the drive toward propping up First World farmers with exports to the Third World makes the developing world that much poorer and that much hungrier.

The United States, the European Union, Canada, Australia, and other rich countries provide some $350 billion a year in subsidies to their farmers. That is seven times what those same nations spend on aid to poorer nations, and it swamps whatever benefit those needy nations might realize from foreign aid. Rich-nation subsidies to their own farmers encourage overproduction and depress prices worldwide. Rich-nation farmers, then, consistently sell their products at a loss and make up the difference in government loan guarantees, target prices, export subsidies, and emergency payments. Oxfam is a London-based international aid organization that focuses on the politics of food and Third World development issues. Its experts figure that 40 percent of an EU farmer's total income comes from government subsidies of one sort or another; 23 percent of the average American farmer's income comes not from the market but from the government.

The continuing government subsidies allow farmers and farm experts habitually to celebrate any innovation represented as increasing production, even as the numbers should tell them that the world cannot absorb—or, at least, pay for—any more. Even the Associated Press failed to note the contradiction of two different articles about American rice growers that it transmitted in early 2002. One story, as published February 3 in the *Wichita Eagle,* headlined "Southeast Missouri's rice renaissance" as great news for farmers in that nearby state who were eager to switch from corn, wheat, or cotton to rice because a new processing mill in the neighborhood made it profitable to do so. The other side was published February 4 in *The New York Times* under the headline "World rice glut hits home." It told the sad story of a global oversupply of rice that led to, among other things, the closure of a mill in a part of Louisiana that had grown rice for decades.

Two years later, *Farm Journal* magazine made both sides of the argument only thirteen pages apart. First, columnist Ken Ferrie advised his gentle readers to give their fields a good squirt of antifungus chemicals every few years, just because. "Don't settle for good yields if you can spray corn with fungicide and harvest great ones." Then, on page fifty-two, economist Bob Utterback recited the eternal lament of rural bean, and seed, counters everywhere: Yields were too good. "It's going to take a whole lot of demand to eat through this inventory," he said. A headline in the August 13, 2004, *Des Moines Register* has probably run in that paper with more frequency than "Yankees win pennant" appears in *The New*

York Times. The headline was "Bin-busting corn crop triggers painful price drop."

The Institute for Agriculture and Trade Policy calculated the real 1998 cost of producing a bushel of wheat in the United States, from seed to export terminal, was not quite $5, but it was being sold for export at $3.50. In a year when U.S. wheat exports added up to more than 28 million metric tons, the U.S. low-balled the world nearly $1 billion on wheat alone. The same thing happens every year with corn, soybeans, cotton, and rice. Such artificially low prices do feed people today but make it harder for them to feed themselves tomorrow.

The World Bank estimates that, for example, the small, cotton-growing African nation of Burkina Faso could better support its agricultural infrastructure and cut widespread poverty there in half within six years if the developed-world cotton subsidies were not cutting the legs out from under Third World farming. In summer 2002, U.N. Secretary-General Kofi Annan urged the wealthy nations to stop subsidizing their farmers so that farmers in the developing world have a chance to compete. The presidents of Mali and Burkina Faso wrote a joint op-ed for *The New York Times*—"Your farm subsidies are strangling us"—pleading for the developed world to reduce or eliminate the annual $5.8 billion in cotton subsidies that were undercutting cotton growers in Africa.

"In the period from 2001 to 2002, America's 25,000 cotton farmers received more subsidies—some $3 billion—than the entire economic output of Burkina Faso, where two million people depend on cotton," wrote Amadou Toumani Tour of Mali and Blaise Compaor of Burkina Faso. "Further, United States subsidies are concentrated on just 10 percent of its cotton farmers. Thus, the payments to about 2,500 relatively well-off farmers has the unintended but nevertheless real effect of impoverishing some 10 million rural poor people in West and Central Africa."

Nations too poor to offer their own farmers subsidies or export assistance could still hope to protect domestic agriculture, and rake in some much-needed cash, through tariffs on imports. But the supposed free-trade regime has pushed the developing countries to lower or eliminate those tariffs. The rich nations have not returned the favor, despite years of promises to do so by national leaders who claim to favor free trade. This state of affairs has led to the development of the Oxfam Double Standards Index. This is a device Oxfam uses to measure the hypocrisy of nations that mock the principles of free trade by demanding access to the markets of poorer nations but leaving in place protective trade barriers such as subsidies and tariffs to protect domestic industries—and, they think, domestic farmers. The EU, with subsidies that have wrecked, among other sectors, the domestic milk trade in Jamaica, tops the list. The second-place United States provides its average corn farmer with an annual subsidy more than fifty times the entire annual household income of his Filipino rival. Canada and Japan round out the top four.

Usually, when a nation allows—or forces—one of its products to be sold in other nations at prices well below the cost of production, as almost all U.S. and EU farm commodities are, the practice is called "dumping." When other nations dump steel or shoes or lumber in U.S. markets, the United States rightly makes a fuss. When U.S.-produced food is dumped on Third World nations, we call it food aid and expect to be thanked for it. Oxfam notes that most periods of American food-aid generosity seem to coincide with periods of low farm prices at home, and so this largesse arguably is inspired more by the desire to bleed off excess stockpiles of grain and boost prices than by any wish to help the hungry. That same aid dries up, Oxfam says, when farm prices rise. In other words, we help nations when they are best able to buy their own food, and refuse to help them when they are in the most need.

Oxfam also figures that rich-nation trade barriers cost poor nations $100 billion a year, more than twice the aid that rich nations give the poor ones. These barriers are particularly harmful to developing nations because they tend to weigh most heavily on the products those nations have to sell—agricultural commodities and labor-intensive goods such as textiles.

More striking than the Oxfam report is one from the USDA—the very agency charged with delivering the checks that prop up American big agriculture to the detriment of the rest of the world. The department's Economic Research Service (ERS) issued a report entitled "Agricultural Policy Reform: The Road Ahead." That report, like Oxfam's, concludes that the trade-distorting aspects of European and American agriculture policies seriously undermine the ability of Third World nations to feed themselves, and thus to become prosperous and stable trading partners. Removing trade barriers and subsidies, the ERS report concluded, would benefit the world to the tune of $56 billion in increased purchasing power annually. About a quarter of that benefit would accrue to the United States, the report said, but Third World nations would also benefit from higher world commodity prices, because they would put money in the pockets of farmers worldwide, and thence within the grasp of everyone else. When 80 percent of the population of many developing nations is still directly involved in agriculture, it is easy to see why America's habit of pursuing a "cheap food policy" is not good for them or for us.

The ERS report estimates that the elimination of all these trade barriers would increase the basic price of food about 12 percent worldwide. Given the minuscule fraction of the average American grocery bill that the food content of products represents, a 12 percent increase in basic commodity prices would hardly be noticed in the United States. It would be a small price to pay for boosting the incomes of farm-based economies around the world, helping to fill their people's bellies and to quiet their sense of deprivation and jealousy.

THE MONEY FAMINE

**Bluntly stated, the problem is not so much
a lack of food as a lack of political will.**

—Jacques Diouf, Director-General,
U.N. Food and Agriculture Organization, 2003

Seldom do members of the Nobel Prize selection committee go
out of their way to name individuals who, despite great academic
stature, will not win because their work deals with how people
live when abstract theories play out in real life. Most unusual is
the person who proves his old critics wrong by winning, despite
earlier predictions to the contrary, the Nobel Memorial Prize in
Economic Science.

Amartya Sen, master of Trinity College at Britain's Cambridge
University, was awarded his field's top prize in 1998 after a shift
in Nobel criteria that recognized values. Previously, economists
won such prizes for producing unfathomable tables of figures
that set out to explain and predict the behavior of money as if it
had a mind of its own, ignoring all human consequences as so
many eraser shavings cluttering up their pretty equations. Ten
years before Sen's award, a member of the Swedish Academy was
quoted as predicting that "Sen will never get the prize." The statement, which was intended, apparently, as a criticism of Sen's
deep soul and wide-ranging mind, instead became a description
of what had been wrong with the Nobel Prize.

Sen's academic and philosophical contributions to human understanding of the world are many, but his primary accomplishment has been to demonstrate that when periods of desperate
hunger have occurred over the past century, they have not generally been caused by a widespread, or even a local, shortage of
food. They have been caused by a shortage of money. Sen opened
his groundbreaking 1981 book, *Poverty and Famines: An Essay on
Entitlement and Deprivation*, with two brief sentences and two key
italicized words: "Starvation is the characteristic of some people
not *having* enough food to eat. It is not the characteristic of there
not *being* enough food to eat."

Poverty and Famines examines the circumstances of the
Ethiopian famine of 1972–1974, ongoing famines in sub-Saharan
Africa, the Bangladesh famine of 1974, and the 1943 disaster of
mass starvation in Sen's native Bengal.

Sen describes how the Great Bengal Famine, which led to the
deaths of between 1.5 million and 3 million people ("Data about
famines are never plentiful"), was in no way caused by a shortage
of food. Bengal set a record for rice production in 1943. Even the
shortage of money, which is what truly causes starvation, was
unevenly distributed across the country. It was a "boom famine,"
the result of wartime inflation and hoarding that priced a great
many goods out of the rural poor's reach. Other factors included
government policies that channeled food to the army and to favored industrial, railroad, and construction workers whose contribution to the war effort was considered vital. Other rules
banned the export of rice even from one Indian province to another and commandeered for military purposes means of transportation that might have allowed food to flow from where it was
still being grown in substantial quantities to where it was
needed.

The wartime footing of Imperial India, with huge public expenditures for roads and military bases, fed a lot of money into the economy. But, Sen explains, most of that money went to workers on military-related projects, and much of it was financed by public debt, all of which created inflationary pressures that farm laborers and rural craftsmen could not withstand. Farmers who owned their own land and even sharecroppers joined the favored urban workers among the ranks of those who survived the Great Bengal Famine. People who starved in the greatest numbers were the landless laborers of rural India, who had no work or whose small wages fell far behind the rate of inflation.

The British government's analysis of the famine, Sen writes, unsurprisingly blamed natural forces and overall food shortages, rather than government policies that made no effort to feed the millions not deemed vital to the war effort or the comfort of the Raj. Blaming imagined shortages, rather than genuine mismanagement and indifference to the plight of the poor, is a common excuse for famine—even, Sen writes, when there is no shortage. In the Bengal case, the official search for a shortage was "a search in a dark room for a black cat which wasn't there."

Similar ironies—if circumstances that kill hundreds of thousands of people can properly be characterized as ironic—existed in Africa and Bangladesh. Crop failures caused by flood and drought did occur, but only in limited areas; these could have been fed by neighboring provinces, even neighboring farms, if crop losses did not also mean the loss of income, the culling of herds that could no longer be fed, and the collapse of land values. Areas known for famine throughout the twentieth century were usually exporting food to somewhere—somewhere, perhaps far away, where people had money to buy it, not locally where people were desperate for food but had no cash.

Although the Nobel committee has recognized Sen's work, there are still many people who do not appreciate its significance, including people who grow food in the United States and Europe and the government officials who set the policies. Enlightened relief efforts are less likely to focus on emergency food distribution and more likely to concentrate on boosting the income of starving families so they can afford to buy food, but the nonacademics—such as farmers, farm groups, and farming bureaucrats in the United States—cling to widespread belief in the Great American Breadbasket, a new version of the White Man's Burden that envisions the Great Plains as the source of sustenance for the brown and black parts of the planet.

Monsanto, one of the U.S.-based corporations that lives to promote the industrial model of agriculture, clings to the food-shortage explanation of the cause of world hunger. In a November 2001 speech to a Washington, D.C., conference sponsored by *Farm Journal* magazine, Monsanto CEO Hendrik A. Verfaillie said, "Low-yielding agriculture is a root cause of poverty, hunger and malnutrition." Verfaillie may indeed be worried about the billions of human beings who live on the edge of starvation, but his worldview of what to do about it is irredeemably colored by the fact that his company makes its money by boosting people's ability to grow, not their ability to buy.

Events in Argentina in late 2001 and early 2002 provide a concentrated, if somewhat extreme, example of how hunger is caused by a lack of money, not a lack of food. The South American nation has long been a giant of agricultural production—grass-fed beef, wheat, and, in recent years, soaring yields of soybeans. Although still considered part of the "developing world" by international economists, Argentina is so productive agriculturally that it is one of three countries—Vietnam and

Thailand are the others—whose food-production statistics are often removed from analyses of the Third World economy lest they skew the results upward.

Yet when an economic crisis moved that nation's government to freeze private bank accounts, the result was unrest that *The New York Times* properly labeled "food riots." The country already suffered from high levels of unemployment, and, the *Times* noted, cutting the middle classes off from their bank accounts made it impossible for them to pay their cleaning ladies, have their shoes shined, or buy anything from street vendors. With no cash in circulation, a succession of governments rapidly rose and fell. Supermarkets—the very symbol of modern plenty—were attacked by rioting mothers. The *Times* described the woes of one working-class suburb of Buenos Aires, quoting a local resident as saying, "Don't they understand that we have nothing to eat out here?"

Nothing to eat? Argentina harvested 16.5 million tons of wheat that year, and was on track to reap another 17 million tons in 2002. The nation was also, according to USDA figures, expected to export 11 million tons, two-thirds of its total output, in the 2001–2002 crop year, most of it to neighboring Brazil and to faraway Iran. Meanwhile, the world as a whole was then carrying a planetary wheat surplus of more than 150 million tons—nine Argentinas worth.

It was not, of course, that Earth at the end of 2001 had 150 million more tons of wheat than all its people could eat. There are still far too many hungry people in the world to say that. Nor did Argentina have 11 million more tons of wheat than could be consumed by its people. The families surrounding the Buenos Aires supermarkets would have been happy to take some of it. It was that Argentina and the world had so *many* millions of tons more wheat than anyone was able to buy. No matter how demanding people get—up to and including rioting in the streets—the human need for food does not automatically translate to the kind of demand economists mean when they talk, as they always do, about the Law of Supply and Demand. The desire for food, moderate or severe, normal or desperate, becomes economic "demand" only when it is accompanied by money. Without money, there is hunger, but it is not the kind of demand—what Sen calls "entitlement"—that will get anybody fed. Or make a profit for any farmer.

The world as a whole has seen its food production more than keep pace with population growth, but individual nations and regions clearly have not. Poor nations fall into a vicious circle because they lack both the wealth to buy food from elsewhere and the capacity to grow enough of it themselves. In fact, they lack the wealth to buy food *because* they can't grow it. The economic security of even the most advanced nations rests to a large degree on its ability to produce food, for domestic consumption, for international trade, or, at best, for both. Nations that have a thriving agricultural base—and those that will develop them in the future—are likely to buy and sell food in the world markets. Even in an economically or politically flawless world, climate and tradition will move food across frontiers, as when a nation that is at the proper latitude to produce wheat or blueberries trades with one suited to grow bananas or coffee. Weather patterns, market forces, and political considerations will always lead to shortages in some places and surpluses in others, the very foundation of trade.

But all that interaction collapses when a nation, a region, or a community has nothing to trade. Through much of human history, a local lack of food has been caused by Nature—drought or flood, disease or infestation. Increasingly, those local shortages could be overcome through trade with other regions that had not been similarly afflicted or that had better planned for the future by storing surpluses from past harvests. And, increasingly, greater food productivity worldwide and the ever-greater technological capacity to store food and move it from one place to another anywhere on the planet mean that local shortages or emergencies are, again, the result of human factors—economic, political, or epidemiological—that are beyond the control of most of those affected.

Constantly increasing food production over the past decades in the United States, the European Union, Australia, and South America has not fed the hungriest of the hungry in Africa or Asia, and that status would not improve even if those developed nations produced still more food at what would be even lower bargain-basement prices. For millions of people living on less than a dollar a day, with corrupt regimes, trade barriers, poor harbors, and nonexistent highways standing between the farmer and the plate—if there is still a farmer and if there is still a plate—hyperproduction of grains and meats is not the answer.

For millions of others, in poverty and not, the glut of food is clearly the problem.

DON'T HELP THAT BEAR

I've lost a hell of a lot more money to
two-legged wolves than to four-legged wolves.

—Gilles Stockton, Montana rancher,
quoted in *High Country News*

And then there's the story about the usually pious man who skipped church one beautiful Sunday morning to go hunting. While he was out tromping in the brush, he rounded a curve on the trail and found himself face to face with a large, hungry bear. Before the man had a chance to step back and raise his rifle, the bear slapped the gun out of the man's hand and growled menacingly. The man turned and ran and soon found a tall tree that appeared to provide his only hope of escape. But as the frightened hunter neared the top of the tree, he looked down to see the bear looking up at him, pondering whether it would be worth his while to climb up in pursuit.

"Dear Lord," the treed hunter announced to the heavens, "I know I have no right to ask for your help in this, my time of trial, seeing as how I did not keep the Sabbath today.

"But, Lord," he continued hopefully, "if you could see your way clear to remember an old friend, well, just don't help that bear."

Those who practice and support the kind of small-scale, independent agriculture that produces an abundance of healthful food in ways that weigh most lightly on the land have been complaining for some time that our government, our institutions of higher learning, our consumers, and, most of all, Wall Street, have been helping the bear. The argument has teeth, and claws, because giant, mechanized, chemical-dependent farms are simply and demonstrably no more efficient or economical than small, independent operations in the United States and around the world. Often, large farms are less efficient per acre, in terms of either the amount of food produced or profits made, than smaller ones. But distorted values—and distorted markets—have made many people, including many farmers, think that in agriculture, bigger is better.

The deliberate distortion of the market toward the industrial model also works in the meat business, as antitrust laws are ignored, pollution standards are riddled with loopholes, and regulatory decisions are made in ways that favor the big and dirty over the small and careful. The claws of that bear cut deep at a small specialty slaughterhouse in Kansas late in 2004.

Creekstone Farms is a small, high-quality meat-processing operation in Arkansas (pronounced Are-Kansas in this case) City, Kansas. It employed some 800 people, a big deal in a town of 12,000, to turn select Black Angus cows into premium cuts of meat, and to do so slowly, carefully, and humanely, by cow-killing standards. Some 40 percent of Creekstone's product was exported to Japan, where citizens like certain cuts of meat and will pay a premium for them.

But when America's first known case of bovine spongiform encephalopathy (BSE)—mad cow disease—was discovered in Washington state at the end of 2003, Japan immediately halted all beef imports from the United States. A condition of reopening its market to American beef was for the United States to test every single cow, as is done in Japan, for BSE. The USDA staff, key members of which had previously been in the employ of the National Cattleman's Beef Association, refused to order such testing, claiming it was both expensive and unnecessary. But the folks who run Creekstone thought it was very necessary indeed. They had lost a huge chunk of their business and, they reasoned, being able to advertise that they test every animal for mad cow disease would boost their product in the eyes of other customers, too.

The USDA stopped Creekstone cold. The processor set up a special lab and hired appropriate experts. But the government is the only legal supplier of the kits needed to conduct the BSE tests, and the USDA was not selling. The official reason was that the test cannot find mad cow in animals less than thirty months old, and because most cows go to their reward at half that age, the tests would not prove positively that the meat was safe. That explanation was probably true, but because the Japanese government only wanted the best effort toward testing, and the Japanese marketers only wanted to be able to put a "BSE-tested" sticker on every package honestly, giving the customers what they wanted seemed the logical thing to do.

What is infinitely more likely is that the USDA, widely seen to be a Stockholm-syndrome happy captive of big agriculture, was doing the bidding of the large meat processors, companies that do not want to test or even to admit that testing is a good idea either scientifically or commercially. Requests from Kansas's Democratic governor and one of its three Republican congressmen to reconsider did not move the USDA. The two Republican senators, Pat Roberts and Sam Brownback, sided with USDA and the big processors, which also operate large plants in Kansas and are generous with their campaign contributions.

Three days before Christmas 2004, Creekstone laid off 150 of its 800 workers. The bear won.

• • •

On the plant side of the business, the fact that the large farms work many, many more acres than small farms can, on paper, make the bottom line swell and the productivity and profitability of large-scale agriculture seem greater. But the USDA, the World Bank, and many other experts have run the numbers and determined that whether the calculation is done on a gross or net basis—before or after the costs of production are subtracted from the value of goods produced—the per-acre productivity of small farms is consistently greater than for larger ones.

In the early 1990s, the USDA put a pencil to it and figured that in the United States, a category of farm with a median size of only twenty-seven acres (a postage stamp by modern standards) could easily post total outputs with a gross value of nearly $1,050 per acre, leaving a profit of $139 per acre after costs. At the other end of the scale, the category with a median farm size of 6,709 acres, grossed only $63 worth of outputs per acre, netting a mere $12-per-acre profit.

If we focus on whether somebody can make a living on a farm, the big-time operation looks better. After all, 6,709 acres x $12/acre = $80,508, more than a decent living. But 27 acres x $139/acre = $5,600, a profitable hobby at best. But if our focus is on feeding the world, the per-acre output of farming operations is the crucial measure, and for that small-scale farms are better. Land, after all, is the one term in the equation that, worldwide, cannot get appreciably larger, and is quite likely to get smaller as good farmland is lost to bad agricultural practices, natural disasters, war, and urban sprawl. Lost or exhausted land can be desperately compensated for by using more machines, more chemicals, and maybe someday exotic biotechnology. But the return on those investments, both to the farmer and to a hungry world, drops precipitously as these tactics are used.

Land widely distributed and managed by thousands of watchful eyes and caring hands returns many times the value of the same number of acres conglomerated into an industrial model. On this basis alone, it should be obvious that the small, independent farm must be nurtured, in the United States and around the world, not for any quaint or nostalgic reasons but because clearly the best way to match demand for food worldwide is to have more of it produced on smaller operations.

Admittedly, the per-acre productivity of the smallest of small farms measured this way is exaggerated, sometimes significantly, by the fact that they may specialize in premium products for a local market—fresh vegetables, say, or flowers—for which there may be limited or seasonal demand. Thus, many of them are not models that can be widely copied. But it is also true that every dollar of net output value is often of far greater worth to the small farmer because the basic costs of supporting his family— the mortgage, taxes and utilities for the farmstead, the truck for going to town, and the fuel it burns—are counted as farming expenses. Indeed, a small farmer's net profit may be gravy, over and above his costs of both farming and living. The profits of a larger operation, however, may be comparatively overstated because it doesn't count the living expenses of any family.

Either way, the smaller farms do so much better than large corporate farms on a per-acre basis because they can survive, even thrive, with the labor of one household or of a family plus seasonal or part-time hired hands. The smaller farms can, and often do, use free manure and compost rather than expensive fertilizers. Unless there is an unusually virulent infestation of insects or weeds, operations small enough for one or two people to oversee can get away with precisely targeted efforts at pest control, weeding by hand, spraying only when insect damage appears, or, better, planting a mix of species that discourage pests more effectively. Large-scale operators, on the other hand, are more prone to take major preventive steps to protect their much larger and more biologically vulnerable investments by carpet bombing large acreages with expensive chemicals—just to make sure.

Even more important, because it is the lesson that can be translated into the most useful practices in the most places, is the fact that smaller farms worldwide are much more likely to make the maximum sustainable use of every square inch of dirt. That is particularly true when the best use of some corner of land is to be left as a wetland or grove of trees, whether for a year or forever. Small freeholders are less likely to squander their resources or their time. The higher annual per-acre output of the small farm is largely due to the fact that small operators are more likely to plant more than one crop a year, alternating nutrient-hogging plants such as corn with soybeans or other legumes that draw necessary nitrogen back into the soil.

Many small farms, especially in the developing world but increasingly in the United States, even plant more than one crop at a time, getting more out of each acre by filling the spaces between rows of wheat or corn with other crops that can be sold, or with a manageable layer of grass or clover that inhibits soil erosion, discourages weeds, and can be used as livestock feed. These old-fashioned farmers not only cut feed costs of their cows, sheep, or goats but are even saved the trouble of moving the feed because the animals are let into the postharvest field to help themselves. The animals pay for their meals, which are less stressful to their digestive systems than corn or other unnatural fodder, by frightening off weeds and insects, churning up the soil, and pooping merrily over a large enough area that nobody has to spend time or money worrying about what to do with all that smelly waste.

The megafarms, on the other hand, plant row crops in huge straight ranks and files, these days even using global positioning systems on their tractors to ensure laser-accurate grids, with large empty spaces between the rows to allow the heavy machinery to pass through. Those tracks, though, expose large areas of soil that are worse than just unproductive empty space. Soil is more likely to wash or blow away and require more chemical pest control. The lack of crop mixing, or even rotation in many cases, demands the increased use of fertilizers—fertilizers that have to be expensive chemicals because, unlike their smaller, manure-spreading brethren, big-time row-crop operations may not have a cow within miles. The cows have all been moved to gigantic feedlots, where the massive amounts of manure constitute an environmental hazard rather than a natural source of plant nutrition.

The fact that small farms more than pull their weight on the production end while doing less damage to the land, giving more people an ownership stake in creating food, and sustaining the economies of rural communities would seem to be enough reason for them to survive in, if not dominate, the world of agriculture. And if the deck were not stacked against them by Washington and Wall Street, chances are high indeed that they would at least survive and even prosper—at least in a modest, one-family-at-a-time way.

To the degree that independent farmers do not prosper, blame must largely be laid not on the farmers but on many years of government action and inaction. The action is years of farm subsidy programs that help make the big farms bigger and the small ones disappear. The inaction is the total lack of interest or even basic awareness on the part of successive Agriculture and Justice departments not only of the need to apply basic antitrust laws to agriculture and food production, but also of the existence of special laws aimed solely at protecting independent farmers from abuse and assimilation by large corporations.

The crop subsidies discussed previously have always been justified not only on the theory that they will ensure a constant and affordable supply of food, to the United States and to the world, but also that they will help keep the family farmer, Thomas Jefferson's ideal, alive and well. The truth, however, is that independent farmers today survive, when they survive, in spite of farm subsidies, not because of them.

To begin with, only about 33 percent of the nation's farm operators get any direct government subsidies. Nearly all those payments go to those who grow the government-anointed program crops, usually wheat, corn, rice, soybeans, and cotton. Operators who raise chicken, hogs, and beef benefit indirectly due to the artificially low cost of grain, mainly corn, used as livestock feed. Growers of just about everything else, from alfalfa to zucchini, may also benefit from government actions—such as water projects—and government inaction—the unstopped and unstoppable flow of illegal aliens to work in fields and processing plants—but not to the extent that either government action or inaction has much effect on their success or failure.

The basic crop subsidies, though, have everything to do with determining the shrinking number of winners and growing number of losers among growers of the basic crops. Because the subsidies paid have always been tied to the amount of crop produced, those who grew the most stuff got the largest payments—or, at least, the owners of the land on which the most stuff was grown got the most money. They would use that money to buy more land, often land previously farmed by lesser-producing neighbors who received smaller government checks, and these new owners invested in more machinery and more fertilizer and pesticides that enabled them to work the added land with no more bodies in the field. Over time, that trend built upon itself and was a significant part of the reason the number of farmers continued to shrink and the amount of land farmed by each survivor—owner, tenant, or corporation—continued to grow.

The Environmental Working Group (EWG) is a Washington watchdog outfit that monitors how the government fails to protect the natural world and, in the case of farm subsidies, actively harms it. The EWG maintains an Internet database of farm subsidies paid, the annual updates of which are not eagerly anticipated among recipients of payments. These farmers often feel they are being characterized as welfare cases, with their reputations besmirched in the eyes of people who do not understand how expensive farming is and how the money they receive goes not into their pockets but to pay for their seed, fuel, fertilizer, pesticides, equipment, and other items that cause money to flow though a local economy that, without them, would be in a world of hurt.

"The payments that have been received over the last several years have been critical to the economic survival of the family farm," Steve Baccus, a Kansas farmer and president of the Kansas Farm Bureau, told the *Salina* (Kansas) *Journal* when the EWG's 2004 update appeared. "The dollars those farmers receive are reinvested in the communities and businesses that would oftentimes wither away if they didn't have a stable local agricultural economy.

"It basically goes right down Main Street."

Yet after seventy years of federal subsidies to agriculture, there are significantly fewer farmers than there used to be, and fewer Main Streets because of it. The bones thrown to smaller farms, as well as the millions that go to larger operations, wind up in the pockets of the big seed, chemical, and equipment companies as much or more than in the tills of small-town banks, restaurants, and department stores. The drill resembles less a rural assistance program than a massive money-laundering scheme. And it is U.S. taxpayers' money that is being laundered.

As documented on the EWG website, the U.S. government paid out $131 billion in agricultural subsidies from 1995 to 2003. While $16 billion of that was for land conservation programs, and $11 billion was counted as "emergency payments" to farmers struck by bad weather and extraordinarily bad market prices, more than $103 billion was for commodity subsidies designed to bolster the value of those crops to farmers even when the market was saturated. But rather than propping up the sympathetic and deserving family farmer, those payments were directed in a maddeningly top-heavy way so that those who produced the most in the past got the most each year.

As the Heritage Foundation report on the subject put it, "Farm subsidies have evolved from a safety net for poor farmers to America's largest corporate welfare program."

The fact that crop subsidies do little for small freeholders and tilt the playing field in favor of the larger operations, which are generally much rougher on the land, is the underlying message of the EWG database. Ken Cook, EWG president, notes that the subsidies are handed out with the idea that they will allow farmers to get by on lower market prices, keeping the price of food low and thus attractive to the international market. But the strategy never plays out that way because no amount of price-lowering subsidies is going to move corn, wheat, and cotton into a market where the people with money are already full and the people who are hungry are broke.

"It's all about placing bets on the international markets that didn't turn out right," Cook said. "Taxpayers have to pay again and again."

Of the $131 billion in subsidy money paid from 1995 to 2003, 23 percent —$30.5 billion—went to a mere 1 percent of the recipients—30,500 of them. Because these operators had the most land and produced the most food and fiber, the 30,500 received the most money, which they could use to buy more land and produce still more of the stuff the marketplace could not afford to buy. The top 10 percent of recipients raked in 72 percent of the subsidies—$94.5 billion. The bottom 80 percent of recipients received only 13 percent of the money, an average of less than $7,000 each, leaving them even more dependent on their ability to make money at a nine-to-five (or maybe three-to-midnight) job in town. Or even more susceptible to being bought out by their larger neighbor.

"The result," said the Heritage Foundation, "is a 'plantation effect' that has already affected America's rice farms, three-quarters of which have been bought out and converted to tenant farms."

That concentration is demonstrated in the names of 2002's three largest subsidy recipients: Riceland Foods Inc. of Arkansas drew $110 million; Producers Rice Mill Inc., also of Arkansas, received $83 million; and Farmers Rice Co-op of Sacramento, California, came in third at a rapid drop-off of $27 million.

Subsidies see to it that the rich get richer, even among recipients that are not obviously in the farm business. In the 1995–2002 stretch, $4.4 million in farm subsidies went to Fortune 500 corporations whose land holdings made them eligible. They included John Hancock Life Insurance Company, which got $2.65 million in seven years; Chevron Oil, $427,000 over the same period; and heavy equipment maker Caterpillar, $320,000. Banking heir David Rockefeller individually gained $518,122 during the same period, and media magnate Ted Turner received $206,948. These are not such huge numbers, perhaps, in the context of all those billions. But they far exceed the median 1995–2002 take of $5,194 per farm operator.

The other reason the rich are getting richer, in both crops and livestock, has less to do with the money the government is spending than the effort it is not expending. The problem is the huge blind spot the federal government, regardless of the political party in office, has when the issue is application of antitrust principles and laws to agriculture. That includes laws and agencies that supposedly target that segment of the economy.

Longtime Iowa State University Professor Neil Harl, who holds both a law degree and a Ph.D. in economics, has the academic chops to advise relevant divisions of the Department of Agriculture, the Department of Justice, and the Federal Trade Commission, along with farm groups and the general public. He is often heard to lament that government lawyers who are supposed to be on the lookout for market-distorting behavior just can't get excited about an industry that still has 2 million active producers, and a market that exhibits no obvious price gouging against end consumers, in this case shoppers in the supermarket.

"One problem in relying on FTC or the Department of Justice is that both agencies seem to believe that agriculture is the last bastion of perfect competition and is competitive by a comfortable margin," Harl said in an address at a farm conference in 2001. "The problem is not one of diminished competition among producers, but among those who supply inputs and process or handle products from the producing subsector."

As children, apparently none of those government lawyers ever played "Monosopy." That's what economists call a situation where, unlike a monopoly that has many customers buying from one or two price-fixing suppliers, a great many suppliers are stuck trying to sell their wares to one or two market-manipulating buyers. Imagine that instead of a small town with a Wal-Mart that has driven all the competition out of business and can now charge higher prices for everything it sells, there is a single Walton-family millionaire living in a town with dozens of independent, and impoverished, shopkeepers whose only hope is to sell something to the town's only real buyer, at whatever price the millionaire is willing to pay.

Because that situation keeps prices low for a farmer's customers—grain processors and meatpackers—and thus at least has the potential of doing the same for consumer prices, government trust-busters simply have shrugged at situations that certainly appear to be against the spirit, if not the letter, of 100 years of antitrust law.

In 1998, for example, a glut of swine in the United States caused the price of hogs on the hoof to plummet by 60 percent, to levels not seen since the Depression. Neither the handful of pork processors nor the increasingly powerful retail chains passed any of those savings along to consumers. The retail price of pork products declined only about 2 percent, and IBP's fourth-quarter profits quadrupled over the previous year.

This blind spot stops the government from acting against consolidation even in businesses where the farmer is the buyer rather than the seller, such as seed companies from which farmers buy and the railroads that ship their goods. The government has remained on the sidelines as four corporations—Monsanto, Novartis, Dow Chemical, and DuPont—have cornered 75 percent of the seed-corn market and 60 percent of the soybean-seed supply. When the Union Pacific and Southern Pacific railroads merged in 1996, leaving the combined line with control of 90 percent of rail capacity west of the Mississippi, the government did nothing to stop it.

On the Monosopy end, giant grain processor Cargill bought Continental Grain in 1998, which gave it 40 percent of the nation's grain-exporting capacity. Despite protests from farm groups and farm-state officials justly worried that this increased concentration in the market would drive down the price of grain at the farm gate or the local elevator, the Justice Department signed off on the deal. It merely required the newly enlarged Cargill to divest itself of nine grain elevators out of the 100 or so it now owned.

At the retail level, fewer and fewer companies control more and more stores. As recently as 1992, there were still enough mom-and-pop, independent, and small-chain grocers that the five largest retail operations together controlled only 19 percent of the market. By 2004 estimates for the top five—Wal-Mart, Kroger, Albertson's, Safeway, and Ahold USA—were as high as 46 percent of the national market and more than 73 percent in the largest metropolitan markets. That gave the retailers enough clout with the processors not only to be choosy about what products they would stock, forcing down wholesale prices, but also to charge the processors for space on the shelves—called "display fees" or "slotting allowances"—which brought the notoriously low-margin retail business a huge boost and in some cases could amount to 50 percent or more of a grocery chain's net profits. As retailers gained increasing market power, suppliers sought more ways to cut their costs—at the expense of producers—and even to seek their own mergers. The 2000 marriage of Pillsbury and General Mills was intended to give the new company more leverage in dealing with retailers, and the 2001 merger of Tyson and

IBP was, according to Tyson's boss, worth "100 feet of shelf space at Wal-Mart."

In 2003, in an action that would seem to violate antitrust laws generally and particularly the law specifically aimed at the livestock business, the Justice Department authorized Smithfield, already the nation's largest pork processor, to buy the pork operations run by the bankrupt Farmland Industries cooperative.

Americans were scandalized to learn in 1921 that the top five corporations controlled 75 percent of the beef-packing business. (Today, four corporations squeeze 80 percent, and few in power seem to care about this greater concentration.) Congress that year passed the Packers and Stockyards Act to prohibit meatpackers from gaining excessive control over the business by banning packer ownership of the parts of the supply chain that led to their slaughterhouse doors—stockyards, railroads, and such— and from taking any actions that would distort the market and give themselves an unfair advantage. The act was extolled by its authors as "the most far-reaching measure" that endowed the secretary of agriculture with "complete inquisitorial, visitorial, supervisory, and regulatory power over the packers, stockyards, and all activities connected herewith." Its first accomplishment was to make dealings at the nation's many stockyards more advantageous for cattlemen and hog farmers by disclosing market prices and making it easier for new buyers to enter the fray. But by midcentury the stockyards were dying out. The new order was a market in which there were fewer buyers and livestock growers were tempted with advance contracts that replaced the risk of the open market with more certain, if depressed, prices.

With the brave new world of protein Monosopy seemingly beyond the government's ability to negotiate, Congress in 1976 made it possible for affected farmers and ranchers to take the processors to court themselves. Their cause célèbre was the case of *Henry Lee Pickett v. Tyson Fresh Meats, Inc.*, a suit brought by ten cattlemen against the nation's largest beef-packer, the since-merged IBP/Tyson, which buys and processes 30 percent of the nation's beef cows. In the case, which was assigned to a federal judge in Alabama, the plaintiffs argued that the growing "captive supply" practice in the industry, in which Tyson also led all competitors, amounted to an illegal manipulation of the market and cost the 30,000 producers who sold cattle to IBP between 1994 and 2002 a total of at least $1.28 billion.

"Captive supply" is a process by which a meatpacker agrees several months ahead of slaughter to pay a producer a particular amount per pound based on a formula set by the processor. It may or may not be the same formula offered to other producers—that's all secret. The practice guarantees the processor a steady flow of live slaughter-weight cattle to keep its production lines humming efficiently without any wasteful downtime. And it promises ranchers and feedlot operators a customer for their finished cattle—at a take-it-or-leave-it price offered by the processor.

That price for live cattle is likely to be bargain-basement, because there are no other buyers bidding up the price. Ranchers who are not part of the system are in an even less powerful bargaining position because contracted cattle provide the processors with nearly all their needs, slashing demand for the ranchers' animals. Because captive supply has the effect of manipulating the market to the disadvantage of the grower, the Packers and Stockyards Act holds the practice illegal.

At least that was how the jury reasoned, and it awarded the plaintiffs $1.28 billion in damages. Amazingly, presiding Judge Lyle E. Strom set aside the verdict, basically buying the Tyson argument that the captive-supply dodge, no matter how much it may disadvantage cattlemen, and regardless of its illegality under federal legislation, must be allowed because, well, that was the way business was done. Tyson presented no evidence that it could not buy sufficient cattle on the cash market, as should occur in a free market, and although this fact convinced the jury, the judge was not swayed.

Another victory for the lie that market-distorting food processors must be given their way in order to keep the supply of food coming.

"People think it's all about efficiency," Professor Harl told *The Des Moines Register*. "It's not. It's all about power, exploitation of market power."

PART TWO

HEALTH

THE RICH GET FATTER, THE POOR GET ... FATTER

Gluttony is not a secret vice.

—Orson Welles

A system that keeps the supply of food high and the price low, even if it costs governments substantial money in subsidies, leads to pollution, and facilitates the concentration of economic power, is easily and often defended as necessary to make food affordable for Americans and families worldwide so they can all be well-fed and healthy.

Reality, however, does not match that ideal. According to Massachusetts-based Worldwatch Institute, a research group looking for ways to make the world's economy just and sustainable, the world's harmfully undernourished people number about the same as its dangerously overnourished people, about 1.1 billion each.

No one should infer from that numerical balance that rich people are eating poor people's lunch. But, as with the orangutan and the chimpanzee, a common ancestor produced our girth and their dearth—a world awash in cheap food.

The waistbands in Americans' pants must be increasingly elastic because the price of basic foodstuffs is not. This lack of what economists call "price elasticity" is also the reason independent farmers in rich nations and rural people in poor ones must cinch their trousers a little tighter, literally or figuratively, all the time.

The basic theory of price elasticity is that if a person is clever, hardworking, or otherwise efficient enough to produce a product at a lower price, then the number of people who will be able and willing to buy that product increases. As the number of buyers increase, the seller can hire more people, buy more equipment, and plan more efficient delivery in ways that make the unit cost of production lower still. Those savings can be reinvested to develop more efficiency and innovation and/or passed along to the consumer in the form of still lower prices, either of which will attract still more customers.

That cycle is usually good for both producers and consumers. It has transformed such items as pocket calculators, computers, VCRs, and cell phones from exotic toys for the rich into affordable, ubiquitous, and even necessary items in our culture, while making those who created and marketed them deservedly wealthy. Yet the cycle is also a justification for all the ruinous aspects of modern agriculture, the excuse for chemical-based, subsidy-dependent, belly-swelling food production and processing: Keep the price down, and more people can buy.

But it doesn't work that way.

Even all those flashy electronic gadgets eventually get stretched to the point on the price-elasticity scale where all the snap has come out of the pants. Once expensive items for scientists and über-nerd graduate students, pocket calculators are now impulse items at the Wal-Mart checkout counter. VCRs became so cheap that electronics firms had to invent DVDs to have something worth selling. Cell phones are now give-away items to lure customers into signing long-term service contracts.

Food is affordable and ubiquitous, at least for most people in developed nations, and a necessity for everyone. It is not, however, elastic. Thus it does not create a great deal of wealth for those who, at the beginning of the multistep value-added process, actually grow the grain or feed the cows.

Even as food marketers launch a constant stream of new products, marketing them mightily with everything from flashy packaging and ad campaigns to loads of added sweeteners to the convenience of premade peanut-butter-and-jelly sandwiches (with the crusts already removed), the basic foodstuffs have long since reached the cell-phone outer limits of price elasticity. That has been the status for a long time, and except for those farmers who can find a niche outside the hyperproduction track, there seems little hope that it will ever be anything different.

The standard economic model of increasing production to lower the cost and make the product more attractive to the consumer is pointless, to the point of self-destruction, for farmers worldwide when basic grain and meat are already so close to free that the cost of food could hardly have less impact on demand than it already does. Most Americans, whether fat and sassy or thin and smart, have little incentive to buy more food than they already do, even if still greater supplies did translate to even lower prices, and those lower prices were indeed passed along the supply chain to the consumer. Poor people, here and around the world, do not benefit from the productivity-boosting farming methods already in use. Nor will they benefit from future developments—the next chemical treatment or a spliced gene—because transportation and trademarks, processing and profits, are what they can't afford, not the cheap grain.

In 1929, American consumers spent $19 billion for food in grocery stores and restaurants. According to the USDA's Economic

Research Service, that was just less than a quarter of the $83.2 billion in total disposable personal income. Total spending on food fell during the ensuing Depression years, to as little as $11.3 billion in 1932. But personal income was falling as well, bottoming out at $45.9 billion in 1933, and thus the percentage of income spent on food remained relatively constant through the Depression. In 1941, as total personal income topped its pre-1929 level by reaching $93.8 billion, spending on food started to climb as well, topping $20 billion by 1942. But food spending as a percentage of income would never return to 1929 levels. In 1941 it dropped below 20 percent and, after a spike in the late 1940s and early 1950s, the percentage of personal income spent on food began a rapid decline. Since 1992 it has remained below 11 percent and more recently has leveled out at just a hair above 10 percent.

In fact, the dime out of every dollar the average American spends for food is not the whole story. Food likely would account for even less than that dime if people were not eating out more and making more use of the relatively expensive prepackaged microwave dinners that busy, two-income, three-child families seem to prefer. This means that not even the whole dime goes for food, in the sense of wheat, corn, beef, pork, and mushrooms. It primarily goes for processing, shipping, packaging, advertising, marketing, retailing, and, in some cases, cooking and washing up afterward—all the activities the USDA calls "food marketing services."

How much of that dime finds its way to the farmer? The answer depends on the type of food, of course, but the more handling it undergoes—processing, cleaning, canning, labeling, shipping—the smaller the percentage of the retail price that stays at the farm gate. A dozen eggs, more likely than many products to have been produced nearby and requiring little improvement other than being nestled into a foam carton, might sell at retail for 91 cents, and more than half of that, 48 cents, is the farmer's cut. Fresh fruits such as lemons and apples also involve little in the way of processing and packaging, but the farm-to-retail spread for them is much larger because there are many more miles to be traveled between tree and teeth. The USDA estimates that the orchard gets 29 cents for the pound of lemons that consumers pay $1.29 for (22 percent) and is paid 19 cents for a pound of red delicious apples that sell for 92 cents (21 percent). The farmer gets 24 percent of the price of that one-pound head of lettuce—18 cents out of the 74-cent retail price.

Percentages of food costs that stay with the farmer drop sharply the more the basic food is changed—called "added value" by economists. The apples in a 25-ounce jar of applesauce are only 16 percent of the retail price, and would be only 18 percent of retail if they instead were in a 64-ounce bottle of apple juice. The tomato grower gets 7 percent of the cost of a size-303 can of whole tomatoes, the same percentage the corn farmer gets from a 1-pound can of corn. A 1-pound loaf of bread that retails for 88 cents returns all of 4 cents (5 percent) to the farmer, and a 1-pound bag of potato chips that cost, by 2000 USDA figures, $3.36 brought the farmer 26 cents.

Is anything wrong with that scenario? Most foods follow a long chain from field to dinner table, and all the players along the way must earn a living. It makes perfect sense that the percentage of the retail price that goes to the original producer would decline as processing and marketing costs increase.

As personal incomes have risen and farm-gate prices for crops and livestock have fallen or remained constant, twenty-first-century American consumers find themselves paying, on average, only about 10 cents out of every dollar of their disposable income for food. The basic commodities—grain, fruit, vegetables, and meat—amount to only 2 cents out of that dime, or 2 percent of consumer spending, with the rest going for processing, transportation, and various middleman markups. Of course, not included in that low price is the cost to taxpayers and to the health care system for crop subsidies, treating polluted water, breathing polluted air, recovering from chemical-induced cancers, and foreign aid for countries whose own farm economies have been wrecked by America's unstoppable surpluses.

If farmers bought less fertilizer, fuel, and pesticide and pumped less water—if their products were driven fewer miles and passed through fewer hands between the farm and the consumer—the basic price of grain, fruit, vegetables, and meat might increase slightly, if for no other reason than that supply might not be so huge as to be in desperate need of a buyer. But even if the farmer's share of the nation's total disposable income did increase—even if it doubled from 2 cents out of every consumer dollar to 4 cents, or even more—there would be benefits for farmers and for the land they no longer would have to work to exhaustion.

The cost to consumers would be negligible. In fact, if there were enough competition among processors and retailers to push those middlemen to absorb the higher commodity costs, rather than pass them along to the consumers, more farmers could cover more of their costs with less taxpayer support, less debt, and less dependence on chemicals, and the added cost to most people could be quite literally nothing. Farmers got 47 percent of the money Americans spent on food in 1950 and even 37 percent as recently as 1980 because of more competition among processors bidding up the price of farm commodities on the one end, and more competition among those processors and among retailers to keep prices down for the consumer on the other.

The cell-phone industry basically gives away the cell phone and sells the service—photos, instant text messages, voice mail, and the ability to chat for hours. Of course, much of that conversation is meaningless, even annoying, to others—and worse, drivers on cell phones can endanger the lives of themselves and others—but such consequences are waved aside and justified on grounds that communication is a basic human need and, besides, it's cheap.

Likewise, the food industry basically gives away the food and sells the service—this in the form of easily eaten sugar-delivery systems, meat and cheese pies ready for the microwave, and on the more health-conscious side, bags of cleaned, mixed greens with a small bag of salad dressing and one-serving-size yogurt cups to eat on the run. Then there are the all-you-can-eat-and-we-do-all-the-work meals in restaurants, which eateries increasingly turn to because rent, labor, and insurance are expensive but the actual food they sell is cheap. Much of the food is unhealthful junk that clogs the arteries and inflates cholesterol and blood-sugar readings—even as growing it depletes land, then enriches a few giant food processors—but these consequences are waved aside and justified on grounds that food is a basic human need and, besides, it's cheap.

Until that first heart attack. Or the diagnosis in a twelve-year-old of what used to be called adult-onset diabetes.

Neil Harl is one of the few experts known to speak out directly against food-sector monopolies and free-market pipe dreams, but he also has inadvertently struck at the heart of what is wrong with agriculture, not only in the United States but around the world. Noting that increased U.S. food exports are constantly championed as the solution to low commodity prices here and hunger in the rest of the world, Harl explained that this export potential remains limited as long as traditional dietary habits and limited incomes hold down both the desire and the ability of the world to buy American food.

"If we could only get the Third World to eat as well as we do," Harl once told an audience of Kansas farmers. The farmers nodded knowingly. But a better response to Harl's wish might have been to ask what the Third World had ever done to deserve such a fate. Surely, it would be more accurate to say that in order for the United States to feed the world profitably, the Third World would have to start eating as *badly* as Americans do. Indeed, there are indications that it is beginning to do just that.

In December 2001, U.S. Surgeon General David Satcher issued a report cataloging the long list of health problems and high rate of premature death in the United States caused by obesity. Every year, the Surgeon General said, 300,000 Americans die prematurely because they are overweight or obese. By 2004 the Centers for Disease Control (CDC) had listed the health effects of being overweight—diabetes, heart disease, stroke, hypertension, depression, even reduced tolerances to the stress of pregnancy or surgery—as being second only to tobacco use as the leading cause of preventable death in America. And the gap was closing fast. Twenty-seven percent of Americans are obese, according to the height-weight standards the government uses. Another 34 percent are merely overweight. Data from 1999 showed that the epidemic reaches American children, with 13 percent of Americans ages six to eleven and 14 percent of those ages twelve to nineteen listed as overweight. Those figures are twice the number of overweight people in the six-to-eleven category and three times the number of overweight teenagers measured twenty years before.

Most important for a discussion of the food supply are data indicating that, rather than leading to skin-and-bones starvation, any but the most extreme poverty leads to obesity, especially for the female half of the population. Poor women of all races and ethnic groups are 50 percent more likely to be obese than are their richer sisters. This is likely due, in large measure, to the fact that poor people have diets heavy in starches and sugars, processed foods that are thought to be cheaper and more filling. Poor women may run to these foods on payday, allowing their children to join them in a high-calorie bender to make up for the two or three days before, when tight money meant they ate very little of anything. These foods are definitely the most heavily advertised, on TV, in print, and in the store.

Not coincidentally, these heavily processed and promoted fat-creators bring the smallest return to the farmer and provide the biggest markup for the processors. Manufacturers buy large quantities of very cheap commodities such as wheat, oats, corn, and, especially, corn sweeteners, and mix them in such endless combinations that they can unveil a new breakfast cereal or snack chip with great frenzy and alarming frequency. More healthful foods such as unprocessed fruits, vegetables, lentils, and beans provide middlemen and retailers with smaller markups and thus receive little of the promotion afforded to sugar cereals, canned meats, and the like.

A few months after the Surgeon General's 2001 report on American obesity, the World Health Organization heard a frightening assessment of the weight of the world. The junk-food habits of Americans are spreading to nations where the lack of food of any kind was previously thought to be the primary problem. Although many children in Africa and Asia remain hungry, the London-based International Obesity Task Force estimates that some parts of Africa have more overnourished children than undernourished ones. (Overnourished, that is, in the sense of too many sugars and starches. A lack of vitamins and minerals afflicts somewhere between 2 billion and 3.5 billion people, according to Worldwatch, a number that includes both the obese and the emaciated.) In parts of Latin America, the study found that up to a quarter of the children ages four to ten were are overweight or obese. The number of overeating-induced cases of diabetes worldwide is expected to double between 1998 and 2025, and more than three-quarters of the new victims will be in the developing world.

By 2004 the U.N. Food and Agriculture Organization found another link between hunger and obesity in developing countries, a link also caused by the increasing availability of American-style cheap, fatty foods. The FAO concluded that when a pregnant woman gets insufficient nutrition, her unborn child's metabolism is programmed to draw as much as it can from the food it does get. But as the child grows up in circumstances offering a little more money and a lot more fatty processed food, they not only eat more but also pack more of it away as fat. Thus, whether children begin life undernourished or overfed, the widespread availability and promotion of junk food wind up making people fat and unhealthy in later life.

There are, of course, starving Africans. There are hungry people on every continent, in every culture. It is not entirely fair to say that Africans are empty because most Americans and Europeans are full, but the truth is that the very economic and political system that keeps us so well fed also deprives others of the chance to feed themselves. The hungry people of rural Africa, Asia, and, most maddeningly, of the rural communities that consistently top the list of America's poorest counties are like the amputated toes of the diabetic or the diseased heart of the cholesterol-loaded person. They are unseen parts damaged by the unthinking efforts of the rest of the body to ensure it is always full.

TO HELL IN
A BUSHEL BASKET

Now there are fields of corn where Troy once was.

—Ovid

If we are what we eat—and if, as we should, we include in that calculation what is eaten by what we eat—then Americans are mostly corn.

Corn itself is healthful enough. The great pre-Columbian civilizations of the Americas were built on what Jared Diamond has called Mexico's crop trinity—corn, beans, and squash. Corn has been well adapted to many climates and has done particularly well in North America, where improving hybrids and increased fertilizer and chemical use boosted yields during the twentieth century from only about 25 bushels an acre to upward of 140 bushels an acre.

Now the most commonly planted cereal grain in the world, corn in its season covers more than 70 million acres of the United States—nearly twice the size of the state of New York—producing more than 10 billion bushels a year. That is nearly five times the amount of wheat produced in this country, and a good deal more than Americans can eat. Half of the corn production is exported; most of the rest is fed to domestic livestock. That prodigious output has depressed the price, of course, to the point that a bushel of corn that costs roughly $3 for the farmer to grow brings only about $2 on what today passes for the open market. But farmers keep growing it because the federal government subsidizes their efforts in the name of a stable supply of cheap food. Under provisions of the 2002 farm bill, corn subsidies will amount to $4 billion a year for ten years.

The urge not to let food go to waste thus combines with the artificially low price of corn to make the crop both a glutton of resources and a food upon which other species gorge themselves. The cattle, hog, and chicken production systems are not directly subsidized by the government in the same way that corn, wheat, and rice are, but because corn is kept cheap, it has encouraged the creation of a factory farm system for beef, pigs, and poultry that would almost certainly not exist otherwise.

The previously discussed shift of chickens from millions of egg-layers to billions of broilers (even if they wind up being deep-fried) was an offshoot of a century of cheap corn—that and the perception that chicken was more healthful than beef (true, if it isn't deep-fried). By 1985, according to the USDA, Americans were eating more chicken than pork, and by 1992 more chicken than beef. But pork and beef production soared, too, largely due to the increasing demand of the fast-food industry. And, again, the ability to meet that demand for meat cheaply was largely made possible by cheap corn.

Cheap corn not only translates to cheap beef, which means we eat more of it and get fat, but it also is likely to be a primary cause of deadly beef, the rare kind that can kill people outright rather than shorten their lives by an indefinite number of years. Cattle are brought off the range, where they have been eating the grass Nature fed their ancestors for millions of years, and are moved into giant feedlots, where they are "finished" on grain, mostly corn. Because the bovine digestive tract evolved over millions of years to handle grass, there has been too little time for cows' digestion to adapt—a mere forty years of stuffing cows with corn cannot create evolutionary change. In short, cows lack an efficient way of processing such starchy food. Some of the corn thus makes it undigested into the animal's lower intestine, where it ferments and turns the cow's hindgut into a highly acidic soup.

A microbe called *Escherichia coli* is commonly found in the intestines of cows and people alike. Like most microscopic organisms, *E. coli* comes in a multitude of mutations, most of them harmless and inconsequential, some actually helpful to the chemical processes that occur in our bodies. But one form, *E. coli* O157:H7, causes a particularly dreadful sort of intestinal bleeding, symptoms of which range from nasty cramps to kidney failure, dementia, and death. In one of those little tricks evolution likes to play on human beings for thinking we are beyond the reach of such primitive processes, *E. coli* O157:H7 loves living in a highly acidic environment. Because the corn-fed cow's unnaturally acidic poop chute kills the other varieties of *E. coli*, including the kind that people can eat three times a day for years and never notice, the nasty strain is the one that survives, in the cow's stomach and in ours.

Even with the unnatural corn diet, deadly *E. coli* is rare. It generally shows up in only 2 percent of the cattle tested. But it takes very little—only ten microscopic bugs—to make a person very sick. Most of it is killed by even the most careless cooking, though it can survive in the core of an underdone hamburger or, more insidious, infect foods not destined for cooking if the manure is used to fertilize plants or if juice from precooked meat gets into the salad bar part of the restaurant buffet (called cross-contamination).

Cornell University scientists published research in 1998 showing that the deadly form of *E. coli* in the meat supply could be practically eliminated if cattle in the last five days before slaughter returned to eating what Nature intended: hay. That didn't happen, apparently because it was too radical a change to the assembly-line process. The Cornell researchers did not mention it, but the science on the subject reinforces the understanding that cows are designed to eat grass, and feeding them anything else is a risk arrogantly taken by greedy people in a search for short-term profit.

Still, even America's hungry cows and chickens cannot eat all that corn. Nor can the nation's families consume that much Kellogg's Corn Flakes or Post Toasties, Fritos, or Tostitos. American farmers also were drowning in cheap corn 200 years ago. Then they turned it into ridiculously inexpensive corn whiskey and turned America into a nation of drunks. Now we turn it into cheap high-fructose corn syrup (HFCS) and are turning the world into a planet of diabetic fatties. The waste-not, want-not result of what might be called the clinically depressed price of corn was the shift of the American soft-drink industry in the 1980s from using sugar to using high-fructose corn syrup. The shift from one sweetener to another, in soda pop and other snack foods, coincided perfectly with the late-twentieth-century epidemics of obesity and Type 2 diabetes.

The growing plague of diabetes, although not contagious the way flu or even AIDS is, is fast becoming one of the most serious, and most expensive, health problems on the planet. In the United States alone, the Centers for Disease Control estimates, 17 million diabetics cost the economy $100 billion in direct and indirect expenses. It is the sixth-leading cause of death and the most common cause of blindness and kidney failure. It can cause complications in pregnancy, exacerbate heart problems, and lead to amputation. With a smart diet and moderate exercise, the nationwide risk of diabetes could be reduced by an estimated 60 percent.

Experts, and not only those in the employ of the corn-sweetener industry, disagree as to whether HFCS is any worse for people than the sugar it has mostly replaced. Some diet experts warn that fructose is worse than sucrose (sugar) because fructose does not trigger the production of insulin, the natural body chemical that assimilates the sweet stuff, or the production of leptin, a chemical that communicates to the rest of the body that appetite is satisfied and eating can cease. These experts do say, however, that HFCS pushes the liver into high gear, which in turn elevates the amounts of bad cholesterol and triglycerides, both of which increase the risk of heart disease.

Other experts disagree, noting that old-fashioned sugar is already half fructose—the other half is glucose—and even high-fructose corn syrup is generally no more than 55 percent fructose.

The hazard, though, is not so much one of quality as of quantity. To say HFCS is no worse than regular sugar is like saying that a bullet from a machine gun is no more dangerous than a bullet from a muzzle-loader. One bullet isn't worse, especially if it isn't Teflon-coated or the kind that explodes. But the number of bullets fired in a brief period from the machine gun is deadly, whereas the single slug from the muzzle-loader might be tolerated or even dodged. Because subsidized corn is so much cheaper than imported sugar that is subject to quotas and tariffs (thank you, Cuban trade embargo), we eat and drink substantially more HFCS than we ever did sugar. HFCS now makes up 10 percent of the calories Americans consume. For some children, that figure is closer to 20 percent. Corn sweetener is so artificially cheap compared with the artificially high price of sugar that the people who make candy and sweet baked goods can increase the size, lower the price, or both to move more of it off the supermarket shelf, through your rotting teeth, and into your bloated stomach.

Soft-drink makers, who basically mix corn sweetener, water, and bubbles and call it good, along with the convenience stores and fast-food restaurants that are their main outlets, simply increase the size of the servings. What McDonald's once called a large soft drink is now a medium, and it also offers an extra large—often with free refills. Big Gulps weren't enough for 7-Eleven customers for long, so the stores soon added Super Big Gulps, offered complete with humongous refillable tankards that commuting drivers balance on their laps in a manner more precarious—though less prone to causing painful injury—than cups of steaming hot coffee. And try convincing a ten-year-old that a 12-ounce can of soda is sufficient when the convenience-store coolers are stuffed with 20-ounce and 1-liter bottles. Busy convenience stores sometimes offer a 1-liter for the same price as the smaller 20-ounce. It's all the same to them.

It's not the same to the palate, however, or even to a person's neural structure, as children raised on a constant flow of cheap sweetener learn to crave it, not just for the taste but also for the emotional lift it can give. The habit of eating high-starch foods and drinking sweetened substances would not be so troublesome, perhaps, except that it displaces the budget, time, and appetite for more healthful fare. The fast-food experience, built on cheap beef served on cheap bread and washed down with cheap soft drinks, pleasantly fills up a person with fat and carbohydrates but leaves no room for fruits, vegetables, and milk. Reportedly, the amount of milk drunk by American teenagers dropped by 36 percent between 1965 and 1996, the same period in which soft-drink consumption soared by 287 percent in boys and 224 percent in girls. U.S. consumption of added sugars increased 28 percent from 1982 to 1997, much of it displacing essential vitamins and minerals.

So if cows cannot, and should not, eat all this corn, and American children cannot, and should not, drink all this corn, how about using it to fuel cars?

For the past twenty years, ethanol has been the promised solution to the dual problems of soaking up the U.S. corn surplus and reducing U.S. dependence on foreign oil. Instead of trading our grain for their oil, as farmers advocated in the 1970s, the new dream is turning grain into gasoline. Ethanol can be made from any number of plant substances, but corn has been the logical source because of the huge surplus. Filling gas tanks with ethanol instead of gasoline—or, more likely, using ethanol mixed with gasoline—is thought to be less polluting and could give America's farmers a few dollars that otherwise would go to Saudi Arabia's sheiks. The fuel source certainly is not a product in short supply and won't be any time soon.

After two decades, however, ethanol constitutes only a fraction of America's energy usage and has yet to become profitable for any producer without the $9 billion in taxpayer subsidies poured into the industry since 1978. Nor is there any particular reason to believe that the billions more in government assistance expected in the coming years will pull the dream into a self-funding reality.

Part of the failure of ethanol to catch on widely can probably be attributed to wary consumers who are unsure about putting a strange product in their gas tank, as well as a corn-distilling infrastructure that never seems to reach a useful economy of scale. In addition, one remaining obstacle has yet to be faced head-on.

Making ethanol requires burning old, dirty, imported petroleum—a lot of it. Corn, after all, is grown with substantial fossil-fuel input: the gasoline in the machines that plant and harvest it; in the trucks that bring the seed, fertilizer, and pesticides and carry the crop to the elevator or directly to the ethanol plant; the energy required to operate the distillery, crushing and fermenting the corn and refining out the small proportion of ethanol in the grain's bulk; the energy needed to truck the ethanol to the place where it is mixed with gasoline, which has its own energy-intensive manufacturing process; and the fuel to deliver the hybrid mixture to the retailer, where customers may turn up their nose at the exotic and, perhaps, more expensive fuel.

In 2001 a Cornell University scientist named David Pimentel calculated that making a gallon of ethanol consumed 70 percent more energy than it produced. Supporters of ethanol and ethanol subsidies criticized Pimentel's figures, arguing that he had miscalculated the amount of energy used and created in the process, from the irrigation pumps that water the corn to the amount of ethanol actually drawn from a given quantity of the plant. But Pimentel logically countered that even if ethanol appeared to be workable on paper, that paper would have ignored what so many other agricultural calculations ignored: the environmental costs of modern high-intensity farming. Those costs include the widespread soil erosion and water depletion involved and the huge amounts of environmentally damaging fertilizer that are used.

Pimentel's solution to the corn problem, if not the energy one, is to stop feeding corn to animals and feed it to people instead. Cattle, particularly, could give up their *E. coli*-encouraging corn diet and return to eating grass, which people don't eat and which doesn't pollute soil, air, and water the way nitrogen fertilizer and feedlots full of cow and pig poop do. He suggests exporting the corn we no longer feed our animals, though he overlooks that part of the reason for the chronic domestic surplus is that the world lacks the money to buy our crops and, if they had it, might not appreciate their domestic markets being flooded with cheap American imports.

Better we grow much less corn, turn the cattle out to eat the grass that will be allowed to grow naturally where all that chemical-assisted corn used to be, pay a few pennies more for everything at the grocery store, and lay up a whole lot more environmental wealth than is achievable with current methods.

DROWNED AND DRAINED

They starved to death. In a storage compartment
full of grain, they starved to death.

—Capt. James T. Kirk,
"The Trouble with Tribbles," *Star Trek*

"Some say the world will end in fire," Robert Frost noted, "some say in ice." Unless the whole of humanity makes major changes in the way it feeds itself, the world is more likely to drown in nutrients.

That doesn't sound so bad. Nutrients. Nutrition. The balanced, nutritious meals that mothers always insisted on, that generations of young women learned about in home economics, and that the federal government itself looks after through the USDA, food stamps, and its food pyramid. How can there possibly be too much nutrition?

Something in the human genome apparently closes our minds to the idea that there can be too much food. Our evolutionary programming, left over from the millions of years spent as hunter-gatherers, drives us to hoard as much food as we can—in grain bins, supermarkets, or subcutaneous stores of body fat—against the almost certain day that supplies will run short.

Our human cleverness, meanwhile, pushes us to produce more and more food, whether we should eat it or not, whether we can sell it or not, again as a hedge against the next natural disaster or man-made catastrophe.

A surplus of nutrients? Even if there were, how dangerous could that be? It certainly lacks the menacing sound of, say, nuclear waste. Yet the waste from nuclear power plants provides the best analogy to the deadly, unintended consequences of modern food production and the global surplus of nutrients that threatens life as we know it. The material that fires either powerful engine of modern convenience becomes a threat to the very people that engine was created to serve. And it remains a threat long after the energy provided by that engine—nuclear fission or agricultural production—has been consumed.

The reason we should fear a planetary flood of nutrients, of course, is that the word means more than just the food humans eat. It includes the many substances we feed to plants and animals, as well as the organic chemicals and compounds that Nature uses in its endless (and mostly balanced) cycle of birth, growth, death, decay, and regeneration. The term includes corn, wheat, and rice; the grasses and weeds consumed by wild and domestic animals; anhydrous ammonia and other synthetic fertilizers; and, of course, manure—Nature's preferred plant nutrient for millions of years.

The obese person packs on the pounds and destroys his health, feeds his waistline and starves his brain, clinging to the notion that as long as he isn't hungry, everything else must be okay. Gluttony in the name of health. Modern agricultural practices churn out tons of food in ways that threaten the economic and physical health of developing nations and of economically deprived families in America. Starvation in the name of nutrition.

The overweight American with clogged arteries, high blood pressure, and diabetes—not the emaciated African—is the image we should keep in mind when considering what government policies should be followed and what consumer choices should be made concerning food. Just as fat, empty calories, and strange substances threaten the health of the individual, overproduction, pollution, and genetic tinkering place our whole system of food production in danger. We feel full, appreciate the effort that went into our meal, and rationalize to ourselves that if a little is good, more is better.

Meat production—beef, pork, and poultry—is increasingly carried out in large-scale feedlots known in the business as confined animal feeding operations, or CAFOs. These Augean facilities are home to thousands of animals, animals that, particularly in the case of hogs, produce more waste per head than even the most incontinent humans. Human communities are expected to collect their own waste and treat it before allowing it to make its way back into the environment. Cities spend millions of local and federal dollars trying to keep up with the challenge. Municipal wastewater treatment plants are basically endless buffet dinners for microorganisms that are invited to consume the nutrient content of human waste before it is released into rivers and streams.

But CAFOs are allowed to handle their waste with open lagoons or other primitive methods that pollute surface and ground water and emit harmful ammonia and other gases—not to mention an unimaginable stink. Often, in vain or cynical attempts to remember the balance of Nature, the effluent of these animal factories is spread on fields, allegedly to fertilize crops—crops that then, supposedly, won't need as much expensive and polluting commercial fertilizer. But the waste is often so concentrated that its nutrient value surpasses what even the most productive grain can absorb and convert to food. That situation particularly holds true in the semiarid areas, such as the Great Plains, where most cattle feedlots and an increasing number of hog lots are located, because corn that lacks enough moisture cannot grow enough to absorb and recycle the nutrients in the concentrated, applied animal waste.

It recalls the question of the comedian who, displaying his shopping cart to the supermarket checkout clerk, asked, "Is this enough toilet paper for this much food?"

The many farmers who are not the beneficiaries of the fertilizer that runs off of feedlots must (or feel they must) ladle manmade fertilizers on their crops. Most often they use anhydrous ammonia, a combination of nitrogen and hydrogen created by a chemical process developed, in two separate steps, by scientists working in the German chemical industry, Fritz Haber and Carl Bosch. Each won a Nobel Prize for his part in developing a process that, without doubt, has done as much as anything in human history to allow food production to keep up with population growth. Experts estimate that without it, Earth's current population of 6 billion people would be no more than 4 billion.

The nitrogen that is necessary to plant growth makes up almost four-fifths of the earth's atmosphere, but it tends to stay there unless split from the air by lightning or metabolized by bacteria that are resident in plants such as alfalfa or soybeans. Before the Haber-Bosch process came along, at the dawn of the twentieth century, fertilizer was manure, supplemented when possible by ground-up human and animal bones. It was enough to keep food growing, but not enough to boost yields to match population growth. After Haber-Bosch, nitrogen fertilizer was much more readily available and able to do its part in the Green Revolution later in the century.

The energizing value of this substance, which farmers buy by the tanker-load, is so high that its first general use was in making bombs. Two million tons of nitrogen are still used in explosives each year—one example being the truck bomb that Timothy

McVeigh used to destroy the Oklahoma City Federal Building in 1995. It is also a key component of the superstimulating drug methamphetamine. Access to this key ingredient, combined with a sparse population that is thought to provide privacy for thieves, is the reason rural states are overrun with illegal, and literally explosive, meth labs.

The concentration of nitrogen in man-made fertilizers is so high, and the efficiency of most grain crops in absorbing it is so low, that most of the nitrogen trickles off into the water table, thence into farmers' wells, city water supplies, and, via the nation's river system, the sea.

The result is pockets of water that are high in nitrates. Some of those pockets occur in local drinking water supplies and, when they are detected, result in warnings that pregnant women and young children should not drink the water. Drinking water with a high nitrate content robs oxygen from the blood, resulting in a condition called "blue baby syndrome" that can be fatal in infants. Rare is the modern farm equipment show that doesn't have at least one booth selling home water filtration systems— selling clean water to people who may well have chosen farm living for its supposedly healthful environment.

Other, larger concentrations of nitrates are increasingly found in the world's rivers and oceans. Nitrates in the water stimulate an explosion in the growth of algae, the green plantlike organisms that often appear as a slimy muck on bodies of water. The algae thrive for a time, then, according to the way of all things, die. Dead algae, like dead anything, proceed to decay, a process that sucks up oxygen from the water around them. Nitrate-fed water systems produce such large populations of algae that when they die, the resulting oxygen loss creates large areas of ocean known as "dead zones." These practically lifeless expanses of once-teeming waters, which ebb and flow with the seasons, are found worldwide, most notably in Maryland's Chesapeake Bay, Russia's Black Sea, Australia's Great Barrier Reef, and the Gulf of Mexico. The Gulf of Mexico dead zone is larger than the state of New Jersey. Its lifelessness is the direct result of the overuse of supposedly life-giving nitrate fertilizers in the farm lands that make up the land drained by the Mississippi River and its tributaries—including the Missouri, Kansas, and Ohio rivers.

Abraham Lincoln knew that control of the Mississippi was key to victory over the Confederacy in the Civil War. He exulted when the fall of Vicksburg meant that "The Father of Waters again goes unvexed to the sea." He could never have imagined that the wondrous river would become a loaded gun aimed right at the heart of the South's once-bountiful shrimp and oyster beds. Our efforts to hypercharge the production of corn, of which we have more than we can reasonably consume, have inadvertently, but directly, crippled our harvest of shellfish and other seafood.

Meanwhile, look at Nature. Earth is lousy with life, all kinds of life, from the poles to the equator. Nobody has to plant it, fertilize it, weed it, breed it, inoculate it, or harvest it. It works as well as it does because the wide variety of plant and animal life has found a balance that, though heartless to individuals and even to whole species, maintains the richness of the biosphere.

When the nutrients that now threaten to overwhelm us are produced, distributed, and consumed in a balanced way, either by Nature or by farmers, they are indeed good things. A balanced ecosystem or sustainable agricultural enterprise has no significant surplus of anything—at least not of anything harmful. In a balanced system, the decaying plants and animal droppings are not problems to be disposed of but are key to the whole system of food production, among human beings as it has always been in the natural world. For thousands of years, and for some operations even today, farmers never needed to buy fertilizer, and herders did not worry about the destination of their animals' excrement. That is because farmers and herders were often one and the same, feeding grain they grew to animals they tended or, more efficiently, simply letting their animals graze the fields at certain times of year, allowing their livestock to return nutrients to the soil in the form of manure.

Before there were farmers, the hooves of many thousands of North American bison or African buffalo served to plow the ground for new stands of grass and brush, while the manure they deposited not only provided the needed fertilizer but also often contained the seed for the next generation of grass. The resulting plant growth provided food for those same herds of herbivores, which in turn served as a constant supply of fresh meat for carnivores, human and otherwise. It was a balance that showed no concern for the individual creature, or the individual species. But the ecosystem was preserved.

Modern agriculture has shortsightedly turned its back on the variety of Nature to impose the uniformity of industry. Most of what is wrong with the way we produce our food flows from the belief that crops and livestock must be as interchangeable as parts in an assembly line.

We breed plants and animals to the point that they, like some inbred royal family, cannot survive, much less flourish, unless they are constantly and expensively protected from the real world. Genetically homogenized life needs heavy doses of irrigation, chemical fertilizers, pesticides, and antibiotics to survive the germs, weeds, and bugs that their more natural ancestors were able to brush aside. The result is food that is more prone to disease in the field and to being the host of food poisoning and unhealthful antibiotic resistance when it reaches our tables.

No living thing truly thrives when it is the only living thing. This is not just an emotional matter for humans and other herding or family-oriented animals. It is a matter of millions of years of evolution for animals, insects, and, especially, plants. The circle of life is not limited to the lion, gazelle, and hyena as depicted in the Disney naturelogs. It gets down to the microscopic level of soil, with microbes, fungi, bacteria, and a great number of slimy, creepy, crawly things that live and die and decay, recycling nutrients, carbon, nitrogen, minerals, and water through a pattern that keeps the overall ecosystem humming with life even as individual creatures and, sometimes, even species pass away. Modern agriculture, rather than having the wisdom to copy this cycle where possible and improve upon it where necessary, has seemingly devoted itself to wiping out this natural hive of life in favor of a sterile setting that supports one organism at a time.

The practice of repeatedly plowing the fields, removing the covering grasses, and poisoning the bugs and the weeds robs the soil of most of its life-giving characteristics. Because this depleted soil cannot trap nitrogen the way living soil can, the farmer needs to pour on chemical fertilizers. Worldwide, the use of nitrogen fertilizer shot up 645 percent in the years 1961–1996. Farmers, who are pushed at every turn to produce more and more, know they are applying more fertilizer than their crops can properly use and, in many cases, are fully aware of the sometimes negative effects that fertilizer is having on ecosystems downstream. Although some farmers will ease back a bit on the nitrogen lever to save the earth and their bankbook, few are willing to risk using appreciably less fertilizer—or less pesticide or less irrigation water. In the aggregate, overproduction undermines prices, but an individual farmer who does anything to limit his farm's productivity will see little immediate benefit. One farmer—or a few or even 100 farmers—cutting back on chemical use will almost certainly have less crop to sell, but the millions of other farmers who did not cut back usage will keep the crop supply high relative to demand. Thus, farmers who practice good stewardship of the earth, or at least their own small corner of it, find they have harmed this year's bottom line.

Notwithstanding all this chemical assistance, soil productivity in the Great Plains decreased 71 percent in the quarter century after it was first plowed. The soil itself, bare between the rows and, during the weeks or months between one harvest and the next planting, naked altogether, easily blows and washes away.

Despite all the efforts of government, science, and individuals since the Dust Bowl of the 1930s, the best topsoil in America— the space where the nutrients are, where the roots can take hold and support both plants and the surrounding ecosystem if people do not pull them up too soon—is being lost seventeen times faster than it takes to be rejuvenated from the decay of living and mineral matter. Raising a bushel of corn essentially costs the landowner as much as five bushels of topsoil. Raising enough corn to add one pound of meat to a cow depletes 100 pounds of soil. These activities increase the need for fertilizer next time, fertilizer that will mostly wash away and poison the world's waters because the weakened earth cannot hold its ground.

The same landscape cannot retain whatever water falls from the sky, so the farmer must irrigate it, in many cases drawing water from underground aquifers that have been there since the Ice Age—fossil water, along with fossil soil, that is consumed as if it were so much fossil fuel. The Ogallala Aquifer lies well beneath the soil, stretching from Texas in the south to the Dakotas in the north. By the early 1990s, pumping water from that great underground lake to quench the thirst of corn and wheat crops and to process cattle and pigs was drawing down the water supply at the rate of three to ten feet a year. It was being replenished by natural processes—rain—at the rate of only a half inch a year.

In sum, the possibility of another Dust Bowl, the kind that darkened the skies across North America in the 1930s, is very real.

WHY RACHEL CARSON
STILL MATTERS

The "control of nature" is
a phrase conceived in arrogance.

—Rachel Carson, *Silent Spring*

People who have not read *Silent Spring* are likely to say that it is a book about pesticides. They might remark, admiringly or critically, that its shocking impact led to a ban on the insecticide DDT in the United States and other developed countries.

People who have read Rachel Carson's 1962 masterwork know that the book is not so much about bad chemicals as it is about foolish people. These readers understand that every serious move to ban a chemical is less a charge against the chemical than it is an indictment of the people who are too bullheaded to use it wisely. *Silent Spring* is not a classic because it lists dangerous chemicals and catalogs endangered species. It is a classic for the same reason other works of literature, from the Bible to Shakespeare to John Updike, are classics. It is a classic because it deals directly with human frailties and hubris.

Because her literary spotlight questioned behavior that was very profitable, at least for a few people, Carson has always drawn a level of hostility experienced by few writers. The loudest attacks came, and still come, from the chemical industry and its flak-catchers. Those attacks continue to this day, and have an audience, because many good people were and are concerned that the alternative to these chemicals—these wonders of modern science—however dangerous they might be, was to risk being overrun by insects that would spread disease, wipe out crops, and otherwise threaten to undo thousands of years of civilization. Even today, as the mosquito-borne scourge of malaria kills thousands in developing countries, those deaths are laid by some directly at the feet of Carson and the government officials who later moved to ban DDT and other chemicals.

What might surprise Carson's attackers, and many others, is that *Silent Spring* did not call for a ban on DDT or any other chemical. Nothing in the book suggested that any of the chemicals whose harmful effects were known or suspected was an inherent evil whose obvious dangers overwhelmed any threat of epidemic or famine.

"It is not my contention that chemical insecticides must never be used," Carson wrote in the second chapter of *Silent Spring*. "I do contend that we have put poisonous and biologically potent chemicals indiscriminately into the hands of persons largely or wholly ignorant of their potentials for harm."

Carson demonstrated how the pesticides of her time were being foolishly misused, especially in the technology-mad United States of the post–World War II era. She explained how people, some of whom should have known better, latched onto the idea of recklessly spraying these killing compounds from airplanes in vain attempts to "eradicate" the latest infestation of whatever insect was frightening people at the moment.

Another of Carson's arguments, which is at least as true today as it was then, is that the primary argument for such sweeping use of bug- and weed-killing poisons is false. The world was not then, and is not now, short of food. If anything, we had a surplus of food that was depressing prices and causing government officials and farmers to seek a remedy. One of the solutions, then and off and on since, was to attempt to reduce the supply by setting aside significant percentages of previously productive farm land. But, then and now, the result of set-asides was an often desperate push by farmers to get maximum production out of whatever acres remained open to them. That maximized production was pursued with two objectives in mind: that enough land—especially other people's land—would be retired to depress supply and increase prices; or, that the amount produced would be part of the formula for calculating federal farm program payments. The former never occurred. The latter almost always did.

Either way, the system provides farmers with a strong incentive to use more pesticides, often covering acre upon acre of land with poisons that may kill some nasty bugs but also will saturate the environment for miles around and for decades to come. Sort of like flooding a city to chase away rats.

In the years since, we have all heard of studies warning of the harmfulness of some chemical, then immediately heard of studies contradicting or deriding the former; the latter studies, not always undertaken by the maker of that chemical, countered the first on the grounds that in order to suffer any of the alarming negative side effects of the particular chemical, a person would practically have to bathe in the stuff. Through much of the 1950s, however, America was practically bathing in DDT, heptachlor, aldrin, and a brew of other chemicals that individuals and governments rushed to embrace. Carson's point was not that there was never a time and place for such chemicals, but that the users of them too often didn't consider the consequences. They knew only that the stuff killed bugs, and bugs are bad, so the chemicals must be good. And if a little of a particular chemical is good, then a lot of it must be better. The fact that those chemicals, used in those ways, also killed birds, house pets, and, she wrote, even a previously healthy horse, was not enough to stop their heavy use until *Silent Spring* caught people's attention.

The stuff we introduce into the environment now, if used properly, might well control some pest or disease that humanity would be better off without and might have minimal side effects. But the twentieth-century habit of drenching pastures with chemicals in order to kill bugs that live only in forests was not just dangerous but profoundly senseless. These chemicals move through the food chain, poisoning creatures we mean no harm, working their way up to us. And even the varmints we are out to get can have their revenge.

The Third World resurgence of malaria, which DDT defenders cling to as proof that such strong chemical killers should not be feared, more likely has been caused by an overuse, not an under-use, of that and other chemicals. Blanketing a Third World country with DDT, which is what some chemical makers would propose, could be folly in the face of the Darwinian rise of a super-mosquito that is immune to DDT-like chemicals and becomes an even greater threat.

Carefully targeted, strictly supervised uses of some chemicals—perhaps even DDT—could be useful, even live-saving. But such practices will work only if they are part of an approach that respects the balance of Nature—the only truly effective means of pest management—and if we refuse to let the poison sellers convince us that a chemical works best when ladled on heavily.

DDT was ladled on so heavily in its day that even though it is now banned in eighty-six countries for agricultural uses, it and chemicals that are likely the product of its decay are found all around the world, in everything from the bark of trees to human breast milk. A blood test of any living human will almost certainly turn up a trace of some pesticide, and multiple species from alligators in Florida to eagles in the Great Lakes to vultures in India exhibit deformities linked to chemical exposure.

In 1948, the beginning of the synthetic pesticide era, farmers and others in the United States worried about insect infestations used about 50 million pounds of the stuff. The bugs nevertheless got about 7 percent of the preharvest crop. By 2000 the United States was applying a billion pounds of chemical bug-killers, yet the insects' take rose to 13 percent. Estimates are that upward of 500 different insect species have evolved an immunity to the most powerful pesticides. And the number of resistant pests is bound to grow, no matter how fast the chemists invent new poisons.

Any attempt to eradicate a variety of pest with a chemical is ultimately doomed to failure. Any attempt to keep up with insect adaptation recalls the *Star Trek* episode about a highly polluted planet where the people had evolved into beings that moved so quickly that they could not be seen by regular humans and registered only as a buzzing sound on the normal ear. When Captain Kirk fired a phaser at one of them, she simply stepped out of the way and watched the poky particle beam go by.

The world has many more bugs than scientists. The bugs come in millions of species and reproduce rapidly and in great numbers. The individual insects that are immune to a chemical will survive its application; those that are vulnerable to it will succumb. The survivors pass that lucky trait along to their millions of progeny, which soon become, in effect, a new species that laughs off the newest terror chemical while munching away at our corn, wheat, and cotton.

When changing Nature, it is crucial to remember that we can never change only one thing. If an insect is wiped out, or if it evolves into a superbug, that change will have an impact on plants and animals around it. A bug that dies out will lead to the local extinction, or outmigration, of the creatures that had relied on it for food, and the plants or insects that the now-destroyed bug had been eating may now multiply unchecked. A bug that thrives, on the other hand, will outcompete other creatures for food and habitat, again altering the local balance of Nature to push aside some species that had once been quite comfortable in the now-changed surroundings.

We—the people—are fewer in number. We reproduce and mature much more slowly. We are not as likely to drop dead immediately upon exposure to pesticides or other dangerous chemicals, but we are almost certainly not going to evolve an immunity to whatever nasty side effects they may cause. Increasingly, cancer appears to be the nasty side effect of a world drenched in pesticides and other such chemicals.

In the years since synthetic pesticides became universally used, the number of cases of certain kinds of cancer have been on the increase—first among farmers and farm workers, then in the rest of the population. The American Cancer Society, the U.S. Centers for Disease Control, and the U.N. World Health Organization are not so eager to draw a connection between pesticides and cancer, though they do note increases in the disease elsewhere as other nations adopt a high-fat Western-style diet. That factor could point to a pesticide link, because fat is where some pesticides accumulate and, years later, express themselves.

Other experts and groups, including the Environmental Illness Society of Canada (EISC), see a clearer link between such chemicals and certain forms of cancer. They include the hormone-related cancers of the testicles in men and the breast in women, both of which are appearing more frequently and at younger ages. Sperm counts in men, especially those who work with farm chemicals, are also significantly down over the past sixty years. Other cancers that appear to have an external chemical link include brain tumors, pancreatic cancer, non-Hodgkins lymphoma, leukemia, sarcomas, and multiple melanoma.

The EISC is one group that also suspects a link between pesticide use and a great many other physical and mental problems, ranging from underdeveloped genitals to childhood behavioral problems. Although symptoms may not show up for years—cancer is frequently caused by an exposure that may have happened twenty years before—unborn fetuses and young children are thought to be the most vulnerable. Their cells are dividing rapidly, their immune systems are not yet fully functional, and their mother-child link is set to screen out toxins that have been part of many thousands of years of evolution, not the latest permutation of synthetic poisons. Children are also more likely to absorb harmful substances while in the womb or while still quite young, when the barrier between their bloodstream and their brain is the most permeable.

It is bad enough for people to suffer the unintended consequences of the ever-greater use of pesticides that by definition are for killing (*-cide* means *killer*). It is quite another for people to be threatened by the unintended consequences of the excessive use of natural—or natural-sounding—hormones and supposedly healing substances such as antibiotics.

Given that parts of the human reproductive system—breasts, testicles, and wombs—have increased rates of cancer, pesticides are not the only likely problem. Another potential culprit is the extra hormones pumped into livestock, mostly cattle and sheep, to make them grow faster (hormones aren't used much for pigs and chickens).

Upward of 80 percent of beef cattle raised in the United States have had hormones injected into their bodies at some time. For a dose that costs about $1, a cattle grower can make a $40 improvement in the final market value of a cow. These chemicals infiltrate the water supply and almost certainly have some strange effects on the development of fish, making males more female and females more male. Dairy cows are also often given a particular genetically altered hormone called rBGH, or recombinant bovine growth hormone, to increase milk production.

Maybe the widespread use of hormones in animals causes cancer in people. Maybe it's the pesticides. Maybe it's both, or neither. Either way, the backstory is the same: Agribusiness wants maximum return on investment in minimal time. A substance that cannot possibly have been tested in every circumstance, in combination with every other substance it might interact with in the real world, is seen or hoped to increase production. Nobody can prove with scientific accuracy that it is harmful, so it goes on the market and becomes so widely used that even farmers who are suspicious of its usefulness or safety must jump on board lest they be left behind.

Food-safety advocates have argued fruitlessly against adding hormones to meat and milk. Both are banned in European Union nations. The dairy-boosting rBGH hormone is not allowed in Canada. But the American regulatory establishment and agribusiness industry, which are basically indistinguishable in their devotion to constantly increasing agricultural output first and desperately trying to find someone to buy it later, have refused to limit the use of added hormones or even to require that food from hormone-treated animals be labeled as such. Dairy and ice cream makers who have tried to label their products "rBGH-free" have faced threats of lawsuits from state and federal officials who argue, with some justification, that there is no test to absolutely confirm that claim, so it cannot be made. These producers had to create more complicated notices, stating only that they seek to be rBGH-free but that the government says there is no scientifically proven reason consumers should care one way or the other.

Opponents have argued that one of the apparent side effects found in rBGH-treated animals, in addition to leg and hoof problems, infertility, and birth defects, is an increased incidence of udder infections. Those infections require additional use of antibiotics to keep the cows healthy; these medicines already are so vastly overused, both in people and in livestock, that they are losing their effectiveness in all species.

Risk of disease, always high whenever animals are crowded as closely together as they usually are in high-intensity agriculture, leads to widespread use of medicine. In the case of factory-processed animals, agribusiness long ago stopped waiting for the animals to get sick before loading them up with antibiotics. So-called subtherapeutic doses—too small to cure a sick animal—have been used since the 1950s to make animals grow faster, giving the processors more meat in less time with less feed. Even when drugs are used to treat an actual illness, growers find it impractical to test each animal to see which ones need treatment, so it is common to treat entire herds or flocks by putting antibiotics in their water or feed.

Even if that process does make the animals bigger, or even cures real diseases, it poses a very real threat to human health. Animals dosed with antibiotics become walking petri dishes, perfect for developing new strains of bacteria that will become immune to the very medicine in the animals. Those bacteria are passed on along the food chain, and when they infect a human being, they already have acquired toughened resistance to many, if not all, of the antibiotics doctors may throw at them.

The growing problem of drug-resistant bacteria is alarming public health officials at all levels. Ailments from childhood ear infections to tuberculosis that were once treated with simple antibiotics are becoming resistant to many varieties of those former wonder drugs. The primary cause of the problem is strictly human. In this "Doctor, can't you give me something" culture, the medical profession is under great pressure to make people feel better immediately by whipping out their magic prescription pads. People demand antibiotics, even when the illness they are suffering from is a cold or flu virus—and viruses are unaffected by antibiotics. With more and more antibiotics in our bloodstreams, the weaker bugs are killed off, leaving all the room and food to the stronger germs, which become fruitful and multiply and won't even flinch when exposed to another dose of the same or a related antibiotic.

The routine use of antibiotics in food production is alarming as well. The Union of Concerned Scientists (UCS) estimates that between 70 and 80 percent of all antibiotics used in the United States goes for nontherapeutic purposes in animals—that is, strictly for growing bigger cows, pigs, and chickens. That amounts to 24.6 million pounds of antibiotics a year, primarily in hogs and chickens and, to a lesser degree, in cattle. Of that amount, says the UCS, 13.5 million pounds are of the kind that the European Union has banned from animal use because of fears they could breed drug-resistant bacteria.

Various entities from the World Health Organization to McDonald's have urged livestock growers to eliminate the use of antibiotics as a growth accelerator, but the U.S. Food and Drug

Administration (FDA) has not been as aggressive as its EU counterparts. Still, the agency has expressed some concern about antibiotics in food animals. An FDA report notes that since 1995, when a particular class of antibiotic called fluoroquinolone was approved for use in poultry to treat or prevent *E. coli* infections, human occurrences of another foodborne bacteria called *Campylobacter* became more difficult to treat because it was resistant to the favored antibiotic for that germ—fluoroquinolone. In 2001, to protect the effectiveness of its variety of fluoroquinolone, sarafloxacin, in humans, Abbott Laboratories sought and received permission from the FDA to remove that drug from use in chickens.

Campylobacter often exists in chicken flocks because it does not sicken chickens, just as *E. coli* often exists in cattle herds because it does not sicken cows. And, just as chickens and cattle are deliberately bred to produce more meat, both bacteria are inadvertently being bred to be ever more resistant to antibiotics. As those antibiotics find their way into food, soil, and waterways, more and more of the planet becomes a breeding ground for diseases that the finest medical minds won't catch up with until, for many people, it is too late.

SOUNDS LIKE SCIENCE

> If you can look into the seeds of time,
> And say which grain will grow and which will not,
> Speak.
>
> —William Shakespeare, *Macbeth*

The scientists who create genetically modified organisms are masters of amazing tricks. It's not only their ability to create a corn variety that oozes its own bug-killer, or a new soybean that can survive the herbicide that is killing all the weeds around it. Those innovations are nothing compared to the semantic adroitness Monsanto and other biotech outfits employ when their audience changes.

When the folks who make, sell, and profit from genetically modified (GM) plants are talking to people who might want to ban, regulate, or so much as label their wares, they make a plausible case that they are merely taking hundreds, if not thousands, of years of plant breeding to the next level. It's nothing so unusual, and certainly not anything so harmful that its production should be regulated in any meaningful way. In addition, the derivative products made from GM corn, soybeans, and whatever else they come up with most certainly don't need any label or consumer warning.

But when the same people from the same company appear before a congressional committee or a court to defend the company's right to patent the GM trait of their product, and to sue for damages when they catch someone using their "intellectual property" without having paid the required license fees and observed the necessary limitations on its use—more crucially, its reuse—the very same substance is so new and so exciting that its innovative owners deserve unique patent protection.

Despite the promises of biotech advocates that their products would feed the world, save the environment, and make rich men and women out of clever farmers—and despite the fears of their opponents that poking the genes of viruses and bacteria into the DNA of grains and vegetables would quickly spread unwanted traits around the globe—the widespread real-world applications of GM foods are so far confined to two technologies. Both have failed so far to live up completely to either the hype or the fear, offering no advantages to consumers and not yet having delivered a better day for farmers. As yet, both have produced profits only to the companies that invented and licensed them.

One of the two widespread uses of biotechnology is a kind of crop that stands up to a particular weed killer, so that when that herbicide is used at a prescribed time, it will kill the weeds but not the crop. That, in theory, can help farmers plan their years and, supposedly, will result in reduced overall use of troublesome herbicides because the guesswork is reduced. Because the herbicide that the GM soybean, canola, and other crops can withstand is Monsanto's Roundup brand, the plants are known as Roundup Ready.

The other GM application being used is a plant that produces its own bug-killer, a version of a naturally occurring toxin called *Bacillus thuringiensis,* or Bt to its friends. And Bt has many friends. For one thing, it works very well on nasty wormy things such as the European corn borer. For another, it is not an artificial substance and thus doesn't count as a chemical pesticide that some farmers, especially those seeking certification as organic farmers, want to avoid.

The promises of these crops—increased yield, lower costs, even environmental friendliness—made them a huge seller among American farmers. From a standing start in 1996, the Bt variety of corn now accounts for 34 percent of all corn grown in the United States, and Bt cotton accounts for 71 percent of that crop in America. Roundup Ready soybeans, also nonexistent less than a decade ago, are now the seed of choice for 75 percent of U.S. soy production. Their popularity is also growing in Argentina and Canada.

So far, though, the fast-growing use of these crops has not been a bonanza for either American farmers or their customers. Research shows that once the higher cost of GM seeds is taken into account, especially when coupled with continued low commodity prices encouraged by chronic overproduction, farmers see little or no improvement in their bottom line. Nor is there reason to believe, in a market with so few middlemen and so many government price supports, that consumers would ever see any cost savings from even the most successful use of GM crops.

The work of Iowa State University farm economist Michael Duffy suggests that by late 2001, genetically modified crops had done little to benefit either farmers or the hungry. "The primary beneficiaries of the first generation biotechnology products," Duffy wrote in a report to the December 2001 meeting of the American Seed Trade Association, "are most likely the seed companies that created the products."

Duffy notes that, in Iowa at least, GM soybeans cost more to plant, less to weed, and produce slightly less per acre than their non-GM cousins. The bottom line comes out about even, at best, for the farmer. Monsanto, though, makes money two ways, first by selling the patented seed, collecting the added technology fees for its GM traits, then by selling the Roundup weed killer, the only herbicide that the tender plants will resist. Bt corn, meanwhile, produces more crop but, Duffy concluded, at a higher cost to the farmer. Only in years when there happen to be a lot of corn borers around, when significant amounts of the crop would otherwise be lost to the little worms, does the use of Bt corn improve a farmer's profitability.

As for benefit to the environment, both Bt corn and Roundup Ready soybeans are supposed to allow farmers to use less of those nasty chemicals that Monsanto, for one, used to make and sell. The problem is, it isn't really working out that way, and is likely to work even less so in the future. It doesn't matter if the pesticide is sprayed less often or being incorporated into the genes of a plant and thus not sprayed at all. Reliance on a single tool for pest control is an imprudent course of action that only stands to backfire.

Information summarized by Charles Benbrook of the Northwest Science and Environmental Policy Center in Sandpoint, Idaho, concludes that Bt and Roundup are like any other pest-control agent—after a while, the pests they are supposed to control develop an immunity to the substance. Along with increased use of Roundup Ready crops has come, not surprisingly, an increased use of Roundup. As the weeds become immune to that chemical, farmers must turn to other, perhaps more hazardous weed killers to keep up. Although pesticide use has been reduced in Bt-type cotton fields, no such reduction has been measured in fields planted with Bt corn. If anything, farmers are now spraying more insecticide on their Bt corn than they ever did on their non-GM variety because of the inevitable increased resistance to Bt among corn borers.

The second development is both frightening and illogical. In truth, although the very mention of European corn borers can cause little beads of sweat to appear on the forehead of a corn farmer, infestations of the little buggers come along once every four years at most. The likelihood of a corn-borer attack is so small that only about 6 or 7 percent of U.S. corn acreage was ever sprayed with corn-borer-targeted poisons in any given year. Generally, only about a third of overall corn acreage is treated with any insecticide, a figure that remained steady throughout the 1990s. Less aggressive application of any poison, as by waiting until pests have been detected or conditions are such that their arrival is to be expected, is the best course of action. It not only saves the farmer money and eases stress on the land, but also postpones the inevitable day when the pesticide no longer works because its targets have developed an immunity.

Even Monsanto—especially Monsanto—knows that the greatest enemy of Bt, or any pesticide, is the evolution of resistant pests. For that reason, it wholeheartedly supports U.S. Environmental Protection Agency rules that require farmers who plant Bt corn also to plant what are called "refuges" of non-Bt corn in the same area—twenty acres of non-Bt corn for every eighty acres of the Bt variety. That way, insects that are still susceptible to Bt will have a place to live and grow, and their genes will still be represented in the general population. Monsanto sales representatives implore farmers to follow those rules to the letter, warning that otherwise the benefits of Bt corn might well be lost. Of course, the farmer thus incurs added costs and troubles, having to plant, maintain, and, for buyers who want non-GM grain, sort two different crops. In years of heavy borer infestation, the non-GM field will take a real licking. Yet in Bt corn, with insecticide implanted in its genes, unavoidable development of Bt-resistant insects is accelerated, and the usefulness of Bt to nearby farmers who never used the GM variety is also threatened. Planting non-GM refuges might slow this process but will not stop it.

Then, there is the cautionary tale about the butterflies.

A 1999 study at Cornell University suggested that pollen from Bt corn kills the caterpillars of Monarch butterflies. Because people who know little about butterflies, less about pollen, and nothing about either naturally occurring or genetically engineered Bt toxins do know that Monarchs are pretty, the story made news and made GM corn look bad. Later studies softened, but never eliminated, the suggestion that Bt corn was not good for helpful or beautiful insects. Of course, Monarch butterflies, like other insects exposed to the modified Bt, might or might not evolve an immunity. The point was not so much that Bt corn harmed a particular kind of insect, but that whatever harm it did pose was even less precise than traditional methods of spraying insecticides.

Scientists and laypeople alike also have valid, if so far unproven, concerns about other possible hazards of GM foods. Moving genes from one organism to another could produce in the receiving plant a protein that induces serious, even fatal, allergic reactions from foods that people were not allergic to before and thus have no incentive to avoid. An alteration that does not harm people nevertheless might hurt wildlife. The use of genes that resist antibiotics to carry other traits scientists want moved from one organism to another can be an additional factor in the dangerous growth of bacteria that cannot be killed by our best medicines.

There is so far no smoking gun to indicate that GM crops make a food dangerous for people to eat. But there are well-grounded fears that genetic characteristics grafted into one plant will be introduced into the broader environment, with utterly unpredictable results. Somehow, the highly skilled scientists who have forced one species to assume some characteristics of another have operated under the assumption that the new genes will stay where they are put. They forgot all about the birds and the bees.

The whole history of life on Earth is the story of genetic drift, traits of one species being crossed with the traits of another to create a third life form. If the new ones are suited to their environment, they not only will thrive but will eventually supplant the previous variety by winning the competition for food and reproduction. In Great Britain, where GM plantings have been legally limited to restricted test plots, a 2002 investigation by *The Times* of London discovered that pollen from GM plants had turned up in the honey of bees that lived some two miles away from the nearest authorized GM plant. In Mexico, the Garden of Eden for corn, GM corn is officially banned for fear that it will contaminate the wide variety of genetic stock that dates back centuries. Yet GM traces have been found in Mexican cornfields, deeply offending a people who take pride in being the birthplace of corn and endangering their chances of selling corn to GM-free countries such as European Union members.

The claim of GM seed makers that they can control their creations is not credible. Part of the reason is that they put undue store in the idea that DNA is destiny—that they need only determine where each gene lives on the double-helix strand, rearrange it to be the way they like it, and send it out into the world in its new, more useful, and "unchanging" shape.

But for scientists to say that they will maintain control of future generations of these plants—and, someday soon, of genetically modified animals—because they have the original code is like architects saying they can build a cathedral that will stand for centuries because they have a blueprint. The structural soundness of a cathedral also depends on the building materials used, the skill of the craftspeople, the stability of the soil, and myriad other factors.

Cell reproduction uses DNA as the pattern for each successive generation. But the pattern is not the building material, not the surroundings, not the various forces that conspire to get grand and small things done. The real building blocks of life, such as proteins and amino acids, are supposed to line up according to the plan set down by DNA. But those materials may well have other ideas or be missing altogether. Or, after a generation or two, the DNA sequence itself could become distorted from its original design and command plants and animals to do things that no experimental geneticist ever hoped to see.

Genes that scientists place in one plant can make their way to other plants, even other species, by the normal birds-and-bees process of cross-pollination. This is something that scientists who work in sealed laboratories seem to have forgotten. A food crop that resists weed killer could turn up in a field where the farmer wanted to plant something else, or simply transfer that immunity to a weed, making it that much more difficult to kill.

GM foods are called "Frankenstein foods" by their many opponents in Europe, and the name may be even more appropriate than even those critics realize. The problem with Victor Frankenstein's creation, after all, was not that he existed, but that he escaped from the laboratory into a world where he had no proper niche, causing both reflexive fear and real damage. Anything "the Modern Prometheus" (the subtitle of Mary Shelley's novel) may have offered humanity was erased when the doctor's experiment went public too soon. Likewise, GM foods have a place in the lab, in the theoretical musings of scientists. If they can stay there, lovingly nurtured by tenured professors who need not be in a hurry to make their inventions pay, the processes involved may yet prove useful to humanity. In the hands of corporations that must pay dividends and satisfy Wall Street with glowing quarterly reports, the pressure to get GM foods out of the lab and into the grocery story is just too great.

Biotech advocates in industry, government, and academia cling to the prospect of GM crops feeding the starving of the world, either by thriving in climates or in soils that discourage all other grains or by providing some special benefit that will compensate for severe nutritional deficits typically suffered by the poor. The poster child for this approach is Golden Rice, so-called because the cellular tinkering this strain of rice goes through adds a yellow tinge to the traditionally white grain. The yellow tint hints at the fact that Swiss scientists figured out a way to take a gene from daffodils, insert it in a particular kind of bacteria called *Agrobacterium tumefacins,* then put the bacteria into rice. The new variety of rice thus produces increased levels of beta-carotene, a chemical that human beings can convert into vitamin A. Lack of vitamin A plagues many of the world's hungry, especially the children. Too little vitamin A makes the body more vulnerable to disease, causes blindness, even kills an estimated 1 million children a year.

The gratification of using a bacteria, an agent that usually makes plants and people sick, to create something that can make everybody well must be great. It certainly brought fame to Golden Rice developer Ingo Potykus, whose face graced the cover of *Time* magazine in 2000. Because Potykus's Swiss Federal Institute of Technology is a government-funded agency, and because the many patents underlying his research might be released by biotech companies seeking good media coverage, the hope has been that Golden Rice not only will make poor people healthy, but also will not stand in the way of their becoming wealthy. The dream is that Golden Rice, and biotech breakthroughs that follow, might be made available to the poor nations of the world for free.

Maybe. Someday. The fact is that for now and the foreseeable future, Golden Rice is the "vaporware" of biotechnology. Vaporware, of course, is a computer-geek term for a kind of software that is always on the horizon, always just about to burst upon the scene, but never shows up in stores or on the Internet. The suspicion is that the rumors of a future product's killer application serve to discourage computer users from switching platforms or buying an existing software package. Buyers wait for the supposedly superior product, even if it exists only in the fevered imaginations of roamers in Internet chat rooms. Meanwhile, rivals who have only real products for sale cannot compete with the ideal expected momentarily from the dominant player in the marketplace.

In the case of food, it means that although more natural, less corporate solutions to world hunger would almost certainly go farther and cost less, GM foods hold center stage in the American popular imagination due to their image as a simple solution to a complicated problem—Golden Rice as silver bullet. This same one-food-feeds-all thinking has led to the Green Revolution's obsession with specially bred strains of corn and rice that take over vast amounts of a nation's farmland, crowding out not only the traditional mixtures of varieties of crops that resist pests and restore the soil, but also the varieties of green vegetables that are a better source of vitamin A and the fish and shellfish that add vital protein to the diet. Lack of a varied diet, not Nature's failure to infuse rice with extra beta-carotene, causes the vitamin A shortages that harm so many.

The most important of the many problems with Golden Rice is that its utility as a vitamin-A delivery system is, so far, much overstated. Current versions of the grain can provide the necessary amounts of vitamin A only if eaten in huge quantities, and even then only if eaten by otherwise well-fed people whose bodies are healthy enough to transform the beta-carotene into vitamin A for the body's use. Sickly children with bloated bellies can eat Golden Rice all day long, and although it may fill them up, it will not provide a healthy, balanced diet—or stave off the deficiency-induced blindness this miracle plant is supposed to prevent. Finally, it has yet to be established that Golden Rice, and whatever GM crops may follow it, will be made available at a price poor farmers can afford, just as it has yet to be shown that people who have eaten white rice for generations will be willing to swallow the yellow kind. After all, the fact that African people eat white corn, and feed yellow corn to their animals, is also a factor in the reluctance of starving African nations to accept yellow corn, GM or otherwise, from the United States.

Opponents of GM foods had their bogeyman as well, and it has been at least the equal of Golden Rice in the battle over popular acceptance of the idea of GM plants. Stephen Jay Gould, the world's foremost evolutionary biologist, called it "pure evil." Nearly everybody else called it the Terminator gene.

The official name for the technology, developed jointly by the USDA and a Mississippi cotton-seed supplier called the Delta and Pine Land Company, was the Technology Protection System (TPS). As the formal name implied, the idea was to protect technology, in this case any sort of genetically engineered trait, from slipping out of the hands of the company that invented it and into the public domain. A plant infused with TPS would grow just like any other such plant. The grain it produced would, in theory, have the same characteristics as any other rice, corn, wheat, soybean, or cotton when it came to growing, milling, processing, or eating. The difference would be that TPS-infused grain would also be sterile. It would not grow another generation of rice, corn, wheat, soybeans, or cotton.

That is why TPS so offends an evolutionary biologist such as Gould, who spent a lifetime studying the way all living things transmit their genetic information on to the next generation, in ways that add to the variety, complexity, and thus the sustainability of life. Sterility, after all, is usually considered a failure of the biological process, not a goal to be desired. Thus, opponents of GM technology gleefully latched onto the Terminator as a prime example of what is wrong with the whole idea of biotechnology. Seeds that cannot be saved from one harvest and planted to produce the next generation of crops not only deprive farmers of a huge part of their livelihood but also short-circuit the process of selecting the seeds that do the best in any given climate and soil type so as to evolve the best gene pool.

The irony, of course, is that Terminator technology could also ease other concerns of farmers and environmentalists, specifically the real fear that modified genes could escape from the crops they were intended for and make their way into the broader gene pool. Genes that make soybeans stand up to weed killers, or help sunflowers grow better in a dry climate, could very well find their way into weeds, weeds that would suddenly be tougher than ever to control. If Terminator technology worked and were used, those concerns might be answered. Of course, to be sold that way, Monsanto and its fellow gene-splicers would have to admit that outside the lab, genes migrate from plant to plant, from species to species. They have good reason to avoid such admission, even if it would justify their investment in the Terminator.

Even more so than Golden Rice, the Terminator exists mostly in press releases and arguments, dreams and nightmares. Little was said of it until Monsanto, already the villain of the piece as far as anti-GM activists were concerned, announced plans to buy, for $1.9 billion, the Delta and Pine Land Company, and with it its patent rights to the Terminator. That announcement in May 1998 armed anti-GM activists with a potent symbol, proof that Monsanto was not out to feed the world but to control the very stuff of life. The ensuing public relations nightmare for Monsanto was all the more remarkable for the fact that in the face of global pressure, Monsanto first announced that the Terminator would stay in the lab for the foreseeable future and then, in December 1999, called off its acquisition of Delta and Pine.

Monsanto thus does not own and, technically, never did own the Terminator. Delta and Pine continues to hold the patent, together with the USDA. As far as is known, the seed named for a cyborg is still an experiment, not a product. But the thought process that led to its creation is now well known and still, perhaps, to be feared. Dr. Frankenstein never was able to rid himself of his creation, either.

Those who argue in favor of widespread use of artificial technologies in food production, and those who have the power to regulate them out of existence but don't out of devotion to keeping the cheap food flowing, demand utterly absolute proof that a proposed technology is harmful before they will consider banning it or even labeling it. Those who demand absolute proof of harm before banning or limiting anything call their irrational standards "sound science." That approach might be better called "sounds like science."

The American contention that GM foods, hormone treatments, and floods of antibiotics have been "proven" safe is so much bilge. By "proven," the government means that it would place the burden of proof on those who would ban or even label GM foods and similar bioengineered products. Absent stark proof of serious harm, the U.S. government insists, such products are to be assumed safe, and countries around the world must import them, no questions asked, or risk reprisals for their protectionist trade policies.

The differing U.S. and EU views on GM crops, as well as on hormone-treated beef and chemical-boosted milk, reflect far more than a simple trade dispute. At issue are the very different ways of approaching the basic task of assessing risks and benefits. And there is no reason the United States should expect the rest of the world to share its view on that important task.

The United States has been fighting the European ban on hormone-treated beef in the World Trade Organization, and threatens to do the same over the issue of GM corn, soybeans, and whatever else comes down the biotech pike as long as the EU maintains its resistance. Such a fight could easily devolve into a full-scale trade war, endangering much more than the $16 billion in U.S.-EU agricultural trade. The U.S. government speaks with one voice in its demands that decisions on the right to sell, even without the buyer's knowledge, current or future GM foods must be based not on the whim of consumer preference but upon the findings of "sound science."

This misnamed theory holds that a new chemical or creation, once subjected to a search for known and suspected dangers—at least those known or suspected by the corporation that invented it and seeks to profit from its commercialization—must be pronounced safe and allowed into international trade without further impediment. Indeed, U.S. scientists, corporate spokespeople, and government officials commonly assert this particular genetically altered substance or that chemical enhancement has been "proven safe."

That is a lie. It has always been a lie. It will always be a lie. Chemicals and organic substances cannot be proved safe, and anyone who says otherwise should not be trusted. The most that can be said is that such substances have not been found to cause particular hazards that scientists were clever enough or suspicious enough to look for. It is not humanly possible to anticipate every use the substance might be put to, every combination of chemicals, temperatures, pressures, and conditions it might be subject to. Thus such a substance cannot be called "harmless." The more remote a substance is to human evolutionary experiences, the more we should doubt any claim that it is "safe."

New substances of unprovable safety, which means most new products, should not reach the market until there has been judgment by an expert that risks may exist but benefits are certain. Even then, precautions generally should be put in place, such as limiting the newest medicines to people who have received a prescription from a doctor, an expert who has concluded that the risk of leaving the disease untreated outweighs the risk of the prescribed medicine. Good health care professionals generally understand that the potions they recommend are not so much cures as poisons with positive side-effects.

Europe's—and honesty's—answer to the myth of "sound science" is called the "precautionary principle." This much superior approach holds that new chemicals and biological processes must be shown to have a clear benefit, outweighing any known or likely risks, before they can be accepted. Applying the precautionary principle to GM foods leads to the reasonable conclusion that it is not necessary to make the case against GM foods stick with firm scientific, conclusive, smoking-gun proof of harm. By this more reasonable way of thinking, it is necessary to make an air-tight case for allowing GM foods before anyone should expect them to be accepted side-by-side with their untinkered cousins. This is especially true if GM foods are going to ride international trade agreements into countries that may not want them, into fields that cannot help but scatter their pollen, into packages that must not be labeled, and into stomachs that are, in very few but very deadly cases, unable to digest them.

What circumstances would constitute an airtight case? Clearly, if corn borers and milkweed were such problems in American and European fields that food supplies were threatened. If milk were so precious it cost $10 a gallon, when people could get it. If meat were so rare that protein deficiencies were epidemic. If anyone on earth were hungry due to a lack of food, then both the real and the still-conjectural risks of GM plants and meat- and milk-boosting hormones could properly be swept aside and those products brought on line with dispatch.

Such shortages do not exist and are not likely to materialize anywhere in the foreseeable future. In addition, the U.S. government's constant harping about "sound science" is ethically and intellectually bankrupt. From these factors, it becomes clear that the United States, even more than Europe or Japan, is following a policy driven by crass economic concerns. The United States has little concern for the scientific and regulatory integrity of its trading partners—we are just mad that they don't want to buy our stuff.

PART THREE

SECURITY

A TARGET-RICH ENVIRONMENT

> For the life of me, I cannot understand why the ter-
> rorists have not attacked our food supply, because
> it is so easy to do.
>
> —Tommy Thompson, Secretary of Health
> and Human Services, 2001–2005

Even though outgoing Cabinet member Tommy Thompson backed off his statement one day after making it, he was right the first time. As head of the giant department that includes the Centers for Disease Control (CDC) and the Food and Drug Administration (FDA), Thompson knew full well that while the world is on the alert for violent attacks by al-Qaeda, most Americans are much more likely to be attacked by *E. coli,* whether or not by terrorists' design.

E. coli, as readers who have been paying attention so far know, is not an international organization. It is a microscopic organism. It does not breed in the poverty-stricken countryside of Afghanistan or the Philippines. It grows in the grain-filled intestines of cattle in Colorado and Nebraska. Just as al-Qaeda struck at cities far from Afghanistan, *E. coli* attacks people who have never been anywhere near a cow's innards.

As occurred on September 11, 2001, when terrorists turned the U.S. transportation system against the nation, this biological threat comes via our highly developed industrial complex. This is the very system that is supposed to feed us in such efficient ways that most people do not give it a second thought—or a first.

In search of industrial scales of efficiency, agribusiness crowds cattle, hogs, and chickens together in the kind of close proximity that can only make a bacterium drool. Often ankle deep in their own manure, animals that people are expected to eat are stuffed with corn or, when that is too expensive, protein feeds that likely contain waste products and ground-up bits of other animals. Feeding animal parts to plant-eating animals is the number-one suspect in the spread of mad cow disease.

The more that modern agriculture has turned its back on the variety that marks the success of Nature, the more it breeds genetically identical strains of wheat and corn. The more it crams uniform breeds of cattle, hogs, and chickens into claustrophobic spaces, the more vulnerable the food chain becomes to the biological risks that occur in Nature. Bacteria, parasites, insects, and other life forms that eat or infect plants and animals are as lazy and opportunistic as any other living thing. If they can find a huge quantity of food in one place, they will hunker down to feast.

Even some people who work within the system of industrialized agriculture, who are not heard to challenge the lie that only a centralized, mechanized system can feed the world, have been quick to note the special vulnerabilities to be found in food production and distribution networks.

"I'm telling you," one expert told *Farm Journal* magazine, "it would be so easy to attack the U.S. agriculture industry that I don't even like to talk about it."

That expert, Charles Beard, should know something about the use of food as a weapon. He not only served as vice president of research and technology for the U.S. Poultry and Egg Association, but he also was a former U.S. Army biological warfare specialist.

"Terrorism in agriculture is not sensationalism," added Iowa State University veterinary professor Harley Moon, "It is reality."

Notable in the discussions about the vulnerability of agriculture to terrorism has been the widespread understanding that the risk is greatly heightened by the concentration and genetic uniformity of food production. More notable, and infinitely more threatening, is the utter lack of reflection on whether that industrial model is a good idea. Most decisionmakers and opinion leaders seem to take it as a given that food is no different from automobiles or computers, the product of a few highly integrated, supply-chain-driven enterprises that operate across the nation or around the world.

We have accepted, on only flimsy evidence, that the efficiencies provided by such centralized operations are the only way to feed ourselves. Suggestions that a more decentralized, less homogenized food supply would be less susceptible to attack, whether by al-Qaeda cells or E. coli cells, are dismissed as naive agrarianism at best and, at worst, an elitist desire to avoid smelly slaughterhouses and make more free-range chicken breasts available to the few who can afford them.

Industrialization, with its large-scale operations, production quotas, and worldwide shipping networks, has led to threats such as mad cow disease and foot-and-mouth disease. The industrial approach to agriculture and unnatural uniformity in the gene pool could lead to real shortages of food when the day comes, as it will, that large parts of the world's crops are decimated by a pathogen that finds them to be its favorite food.

Massive numbers of animals and plant products are crammed through narrow choke points before being distributed across the country and around the world. Millions of acres of genetically identical crops are grown in machine-straight rows, easy to plant, to cultivate, to harvest, and, if you are a so-inclined terrorist, insect, or microorganism, easy to poison, devour, or infect. Staggering numbers of hogs and cattle are kept in huge feedlots, near one another and surrounding the slaughterhouses to which, and in which, they will shortly be dispatched, all in the name of industrial efficiency.

The vast majority of America's beef, for example, is processed in less than a dozen plants owned by only three corporations—Swift, IBP, and Excel. Livestock of all sorts are shipped great distances from birth to feeding to slaughter, spending the last part of their lives (cattle) or all of them (pigs and chickens) crammed cheek by jowl, or beak by giblet, into horrendously foul, smelly barns and stockyards, boxcars and trucks. Then, once hogs and cows and chickens are turned into pork chops and T-bones and nuggets, industrial concentration quickly becomes industrial distribution. Refrigerated trucks carry precut meat to supermarkets and restaurants all over the country. An old saying attributed at different times, in slightly different forms, to Mark Twain and Winston Churchill holds that a lie can be halfway around the world before the truth gets its boots on. It is no allegory, but quite literally true, that a large batch of contaminated meat can be halfway across the country, mixed into hamburger in hundreds of restaurants and supermarkets, before the U.S. meat-inspection system finds its clipboard.

Meat is recut, reground, and repackaged, in the process losing its original labeling and enhancing difficulty in tracing should there be a recall. The vast majority of the meat recalled in any action is never physically returned. The delay between detection and action means that two-thirds or more of any batch of suspect beef will already have been eaten—most, by luck or by good cooking, without having harmed anyone. But the low level of returns means a low level of follow-up testing, and with little demand for refunds, this leaves little incentive to avoid future recalls.

The opportunity is there for a terrorist—or a disgruntled slaughterhouse worker or just a deranged individual—to introduce some germ or virus or fungus into the food supply and have that microscopic bomb carried for him to homes and restaurants and ballparks and picnics across the country. Such a person might have to try more than once before his attack would get through government and industrial safety systems that are in place. But the individual could try more than once because, unlike the terrorists who use fully loaded jetliners as weapons, a first attempt is neither public nor suicidal.

When terrorists can evade the Immigration and Naturalization Service's watch lists, they can commandeer a jetliner and fly it right down the gut of the world's greatest city. That is what al-Qaeda did in 2001, and, we are told, the entire government is being overhauled to ensure it does not happen again.

And when pathogens can evade the Food Safety and Inspection Service's "critical control points," they can ride the multistate supply chain right into the guts of thousands of children. That's what *E. coli* does every year and, say the regulators and the regulated, nothing is really wrong.

In summer 2002, ConAgra, one of the largest food processors in the world, recalled nearly 19 million pounds of ground beef that had been processed at a single plant in Greeley, Colorado. The company took this action at the USDA's suggestion—a suggestion only, because although the government has the power to recall the electric frying pan the beef might be cooked in, the USDA does not have the power to order such a recall of tainted meat—and won't have this power until Congress summons the nerve to face down the meatpacking industry.

This particular outbreak of *E. coli* struck at least forty-eight people in twenty-three states, and one death was attributed to the bacteria. On average, *E. coli* kills sixty-one Americans every year and poisons 73,000. "Poisons" is a better word here than the common euphemism "sickens." Even people who survive *E. coli* have been in far worse shape than just "sick."

One of the flaws of daily journalism is that the compression of time and space leads reporters to use short, simple words to express more complicated concepts. Sometimes, though, journalists have taken time to explain how "sickened" someone struck by *E. coli* can be.

Articles in *The Washington Post* and *The Denver Post* and passages in Eric Schlosser's *Fast Food Nation*, among others, describe the torture of *E. coli* food poisoning. Stomach cramps from hell. Toilet bowls filled with blood. Holes drilled in skulls to relieve pressure on the brain. Loss of kidney function, perhaps permanently. Young children who progress from fear, not understanding why they are being punished in this way, to despair, not understanding why they are being denied water, to dementia, not even recognizing the helpless parents who stand by their bedside, to death. Children and the elderly are the most vulnerable to the ravages of food poisoning, of course, due to weaker immune systems.

"I told my son to tell them to let me die," seventy-year-old Velma Wagner told *The Denver Post* after recovering from food poisoning brought on by eating ConAgra beef she bought at Safeway. "I wouldn't wish this on my worst enemy."

Fast Food Nation tells the heartbreaking story of six-year-old

Alex Donley. He took ill on a Tuesday night after having eaten a tainted hamburger; he was dead by Sunday afternoon. In the meantime, he suffered in ways no human being should ever have to suffer, and in ways no parent should ever have to witness. Holes drilled in his skull. Tubes inserted in his chest. Internal organs being eaten alive by Shiga toxins. Hallucinations. Dementia.

"The sheer brutality of his death was horrifying," said his mother.

Following up on the 2002 ConAgra recall, *The Denver Post* reported that health officials in two other states, Utah and Oklahoma, were trying to determine if beef from the suspect lots had made its way into their states, but that there was no rule or process that allowed them to get that information.

Testing as early as February of that year indicated tainted meat was coming out of the Greeley slaughterhouse and moving through smaller processing plants, where beef from various sources is often mixed together, and on down the chain to Safeway and Kroger stores, restaurants, and other customers. But the discovery of *E. coli* at a Montana processor was not even reported to ConAgra, much less the rest of America, even though tests indicated that the Greeley plant was indeed the source of the contamination. The USDA later explained that it issued no warning because it was not sure of the source.

When more samples of ConAgra product, this time at a Denver processor, were also discovered to be carrying *E. coli*, the company was notified, and it issued a recall of 354,200 pounds of ground beef on June 30—thirteen days after the first Denver test showed *E. coli* in ConAgra beef. In the meantime, the *Post* reported, Safeway stores in Colorado, unaware of the danger, were running a two-for-one sale on ground beef it got from ConAgra. As more tests started coming back positive, and people started getting sick, ConAgra expanded the recall to nearly 19 million pounds of ground beef on July 19.

It was the second-largest beef recall in history, trailing only the 1997 recall of 25 million pounds of beef by Hudson Foods. Hudson survived only a couple of weeks after that recall, dying by the very industrial mind-set it had lived by: It was swallowed whole by the ever-larger Tyson Foods. Another food giant, Sara Lee, pleaded guilty to selling *Listeria*-tainted meat in 1998—and walked away with a $4.4 million fine, a pittance compared with its annual sales of $17.7 billion. Fifteen deaths and six miscarriages were associated with the outbreak. ConAgra survived its 2002 troubles as well, though its $1.4 billion plan to spin off its fresh beef and pork business to a new holding company was delayed until the controversy blew over.

The pain that could be inflicted by a deliberate attack on the food chain is as unimaginable now as the destruction of the World Trade Center towers was before 9/11. But as long as agriculture tolerates this overdeveloped, overcentralized, target-rich environment, it is only a matter of time before we suffer a massive outbreak of food poisoning—whether or not someone meant it.

Then, perhaps, members of Congress will be held to account for refusing to give the USDA the power to recall meat or shut down processing plants it has reason to mistrust. One such bill (which, at this writing, was languishing on Capitol Hill) is called the Meat and Poultry Pathogen Reduction and Enforcement Act of 2003—even though it's now 2005, which shows how long its backers have been working on it. The proposal is known to its friends, which should be every American, as Kevin's Law. It is named for two-year-old Kevin Kowalcyk, of Madison, Wisconsin, who died in the summer of 2001, most likely of untreatable *E. coli* poisoning, after days of excruciating misery for him and his parents. The child had to be strapped down to receive kidney dialysis. He crawled around in great pain, vomited bile, and sucked the water from a washcloth when doctors could not make him understand that drinking any liquid would worsen his condition.

If America-hating terrorists, the kind that blow up giant buildings and behead defenseless civilians, do not deliberately use this poison against us, it will probably be because even they have their limits. The government is certainly doing nothing to stop it.

Nothing.

The reasons for not passing Kevin's Law, or similar proposals that have come before it, have always been weak mumblings in Congress about the need to protect jobs in the beef-packing industry. Of course, that assumes anybody in power cares about the people who work in the Western world's most disgusting enterprise, or about keeping the lines moving at top speed so as to keep the consumer's price of meat down. In fact, regular pathogen testing, which some restaurant chains already require, adds only 2 cents a pound in cost, and the wholesale price is set by what the constricting number of supermarkets will pay, not what the meat processors choose to charge.

Other laws named after their young human victims—Amber Alert, Megan's Law—or cute dog victims—Scruffy's Law—sometimes take a while to get passed, but in the end, there is no violent pedophile lobby or pro-abuse-of-animals PAC to fight them off. American agribusiness, on the other hand, is one of Washington's strongest lobbies, donating more than $140 million to presidential and congressional candidates since 2000.

About the time that ConAgra was recalling its potentially poisoned hamburger, the General Accounting Office (GAO) was issuing a draft report attacking the USDA's meat-inspection program as fatally flawed. Industry, said the investigative arm of Congress, has been entrusted with too much of the inspection of beef, pork, and poultry, while a smaller number of government inspectors were supposedly free to roam the plants, oversee operations, and apply scientific tests for pathogens rather than rely upon the old, repetitive, stress-inducing practice of "poke and sniff."

The poke-and-sniff test clearly had its limitations. *E. coli*, for example, does not make a cut of beef look or smell funny in any way. A relic of the day when meat contamination meant rat droppings and sawdust, the individual inspection of carcasses was likely to miss the microscopic carriers of illness and death. Those limitations became horribly apparent in 1993, in what became known as the "Jack in the Box outbreak." Four children were killed by tainted beef from those fast-food restaurants, and more than 700 people in four states were stricken. The political response, three years later, was the USDA's Pathogen Reduction/HACCP rule. HACCP stands for Hazard Analysis Critical Control Point, though Eric Schlosser quoted doubtful federal inspectors as saying it really stands for "Have a Cup of Coffee and Pray." Under this rule, instead of federal inspectors peering into each brisket, belly, and gizzard, the processors would be responsible for identifying key points along their production lines where poisons were likely to enter the process. The plants' testing and the USDA's oversight would include laboratory testing for those hazards, not just reliance on an experienced inspector's knowing nose.

In theory, that practice makes a lot of sense. In practice, well, Ralph Nader's *Public Citizen* labeled it "Don't look, don't find." That was backed up by the GAO's 2002 findings that the scientific work of examining, testing, retesting, and correcting simply was not being done. Inspectors were not properly trained in the more complicated techniques, and slaughterhouses were allowed to get away with far too many failed tests before any action was taken.

The fact that the *E. coli*–tainted beef escaped the ConAgra plant in Greeley, only to be discovered later at smaller resellers, backed up the GAO's conclusions. The USDA promised improvements, specifically more and better training of inspectors, improved oversight of packers, and better statistical sampling.

But the consumer advocates, wisely, are not buying these assurances. The USDA's rival missions of inspecting the products of American agriculture and promoting their use, at home and abroad, throw into question the seriousness of inspections and the likelihood of bans and recalls. So, too, does the very nature of the meat-processing system, a behemoth of sweat, sinew, and shit that quite literally breeds poison.

It is hard to imagine that any inspection regime, no matter how empowered by the political system, could keep up with not only the size but also the speed of the modern slaughterhouse. As many as 200,000 cattle may be held in a feedlot, while up to 400 cattle an hour are slaughtered at some of today's larger processing plants. The speed at which the animals are pushed through the plant pressures employees to keep up, leaving them unable to follow even the company's own policies for sanitizing equipment or preventing hunks of manure from falling off the hide or flowing out of the freshly eviscerated intestines. It is in the manure, and nowhere else, that *E. coli* lives and breeds. As long as it stays there, not even the cow gets sick. But given a big operation's size and speed, the minimal standard of keeping cow poop out of the hamburger is apparently more than the most up-to-date slaughterhouse can handle.

Inability to keep manure out of meat does not breed confidence that high-speed slaughterhouses will be able to keep brain and nervous tissue apart from the beef, either. That has the potential to be particularly tragic, as eating such tissue is the only known way that the brain-wasting mad cow disease can make its way from cattle to humans, where it becomes the grim, always fatal brain-wasting malady called variant Creutzfelt-Jakob disease. It first entered cattle when agribusiness, ignorant or unconcerned about the fact that cattle are not carnivores, and certainly not cannibals, decided it was efficient and cheap to feed cattle ground-up bits of other cattle, including brain and nerve bits that are the only known carrier of the disease.

There may yet be some hope in the fact that food safety has taken on a certain Defense-of-the-Realm tenor that it clearly lacked before. With a strong and wealthy lobby battling any more government influence in their business, the giant agribusiness corporations have long managed to keep the number of inspectors far below that necessary to truly ensure a safe supply of meat and poultry, and to resist all attempts to allow the USDA to recall suspect meat in the same way that other government agencies can recall exploding cars or flammable baby sleepers.

THE AXIS OF WEEVIL

No man chooses evil because it is evil; he only mis-
takes it for happiness, which is the good he seeks.

—Mary Wollstonecraft Shelley,
author of *Frankenstein*

Late in February 2002, in a large room in the Virginia suburbs of
Washington, D.C., U.S. government experts and a few representatives of allied nations gathered to lament the continued resistance of an intractable foe, one that refuses to see reason or play
by the rules of international behavior. Those stating the case for
overcoming the defiance of this opponent included people from
the very society being targeted, people who could only express
exasperation at their culture's uncooperative attitude.

The meeting was not at the Pentagon or CIA headquarters. It
had nothing to do with Iraq, Iran, or North Korea. Instead of the
Axis of Evil, the discussion sponsored by the USDA was about the
nations that still resist the cultivation and import of genetically
modified crops. China, Africa, and of primary concern to this discussion, nations of the European Union apparently form a sort of
Axis of Weevil.

These nations have refused to open their ports, their fields,
and their mouths to American GM foods and instead have been
placing the wishes of their own citizens and the concerns of their
own scientists and farmers ahead of the demands of outsiders,

often refusing to approve the use—or, in the case of China, merely the import—of this new biotechnology.

It was more than the panel—a British economist, scientists from Finland, Brazil, and Kenya, and a U.S. State Department official—could grasp. Panel moderator Julian Morris, a research fellow at the Institute of Economic Affairs in London, opened the discussion with a description of the poverty and hunger found in India. He left it to the audience of more than 1,000 people—most of them USDA employees—to conclude that resistance to American-made GM foods in Brussels and Beijing is part of the problem in Bombay. Helena Von Troil, of the Nordic Committee on Bioethics, apologized for her continent's failure to bulldoze all opposition to GM crops. She said acceptance would come, but given the poor job European GM-food advocates have done, it would probably take another ten years.

Alan Larson, U.S. undersecretary of state for economic, business, and agricultural affairs, said the world could not wait that long. Larson argued that the kind of restrictions the EU seeks to place on GM foods, including rules for labeling, liability, and tracking, constitute "a very serious threat to the integrity of our agricultural trade system."

The panel, part of the USDA's annual Agricultural Outlook Forum, was educational, but probably not in the way its organizers intended. The one-sided discussion was a perfect illustration of the unquestioning support for GM crops that is prevalent among U.S. government officials and economists. It demonstrated the unreasoning arrogance of those who have embraced GM crops as a solution to hunger in a world already so awash in low-priced grain that huge government subsidies are necessary to keep American and European farmers afloat.

The U.S. government has been squarely aligned in the pro-GM

camp from the beginning. In 1999, President Clinton awarded the National Medal of Technology to the Monsanto scientists who invented worm-killing corn and pesticide-proof soybeans. Agriculture and trade officials in the Clinton and Bush administrations have made it a priority, though not always a successful one, to fight for the right of U.S. biotech companies, farmers, and ranchers to export genetically modified seeds, crops, and products, as well as beef and milk that come from hormone-enhanced cows.

Toward the end of the Clinton years, Agriculture Secretary Dan Glickman started to make noises that indicated he doubted the divinity of GM foods. The evolution of his position was slow and often not noticed, but Glickman suggested it was beginning to look like a big waste of economic and diplomatic capital to try to push entire continents to buy something they clearly did not want and did not need. There was never a major policy statement to this effect by him or anyone else in the Clinton administration. In fact, in at least one 1999 speech, Glickman included his "the customer is always right" ideas. But those passages were not part of the official printed version of the speech released by the USDA.

No such soul-searching has been evident in the George W. Bush administration. Ann Veneman, Bush's first USDA chief, constantly sounded the call of "sound science," especially in attacking Europe's resistance to GM and hormone-laced food. With Glickman spending time at Harvard, on his way to becoming president of the Motion Picture Association of America, the U.S. government's claim continues to be that all the science that has been done on the matter points to GM foods and hormone-boosted meat and milk as being perfectly safe for consumers. Therefore, the U.S. argument goes, any resistance to importing them must be a purely economic, purely protectionist action on the part of unreceptive nations. If French or African officials move to ban the import of GM foods, the American argument goes, the only real reason they have for doing so is to protect their wheat and rice farmers from competing with the products of their American counterparts.

U.S. agriculture and trade officials attack even the European, Japanese, and Australian requirements that GM crops, or the corn flakes and tofu made from GM crops, or the beef and pork that were fed GM grain, be labeled as such; these requirements are portrayed as somehow being in violation of the spirit and letter of international free-trade agreements. According to the U.S. government, a small sticker with the letters "GM" on a box of cereal or package of hamburger constitutes a trade barrier.

Politicians being what they are, and French and Japanese farmers being as politically powerful as they are, there is truth in the charge that anti-GM policies are driven in no small part by the desire to keep out foreign competition. Japan and the EU have long had other policies that seek to support their farmers and keep them in business, not only for crass political reasons but also because their citizens know, unlike Americans, what it is like to be hungry. All that post–World War II aid—the Marshall Plan in Europe and General MacArthur's regency in Japan—not only fed people but kept entire nations from collapsing into chaos and/or communism as they rebuilt from the ashes of war. That regime of aid in those nations quickly gave way to their reinvigorated desire to be as self-reliant as possible, especially in basic necessities such as food.

The people and governments of Europe and Japan quite reasonably want their food, as much as possible, to come from their own fields. The food people eat is more than a consumable item like paper or toothpaste. It has cultural significance and is a matter of national pride, around the world as much as it is in America. The U.S. government's apparently willful blindness to that fact can only hurt our credibility in whatever international forums it would use to have its way.

Further proof of this sad fact lies in analysis of the U.S.-China feud over GM foods. The panel discussion that opened this chapter included China in its Axis of Weevil, not because China does not create, grow, and eat GM foods, but because it does not import, grow, and eat *American* GM foods. The Chinese countryside is teeming with GM crops, and even more are in Chinese laboratories. The January 2002 issue of *Nature* quoted Chinese research institutes' claims to be working on 141 different varieties of GM crops, sixty-five of which have been approved for use by the nation's farmers. The $112 million China was reported spending annually on GM research is dwarfed by the American devotion of $2 billion to $3 billion a year, but it also far exceeds the estimated $15 million a year spent by even leading agriculturally developing nations such as India and Brazil.

Chinese GM research, unlike its American counterpart, is strictly a government operation. Whereas American biotech efforts have so far turned out stuff that is fed mostly to animals—corn and soybeans—Chinese scientists have put their efforts into food for their burgeoning human population—wheat, rice, potatoes, and peanuts. Given that China has so many people and so little good farmland, the nation is a logical candidate to throw in its lot with biotech. Carefully done, such pursuits in China could even be deemed in keeping with the precautionary principle of need preceding the creation of a product. Having 141 new plants in the lab, with sixty-five of them in the field, hardly sounds precautionary, though, and a nation that is already pressing the limit of its food production could easily find itself in deep trouble if any of its many experimental plants transforms into the worst weed the world has ever seen.

The United States, meanwhile, continues to wring its hands over the way China, while developing GM crops of every shape and size, sounds positively European when it comes to importing GM crops from other countries. U.S. agriculture and trade officials complain that Chinese import restrictions and labeling requirements stand to destroy what had been a $1 billion export market for U.S. farmers. If it wants to be taken seriously as a full partner in international trade, China cannot admit that its import restrictions on GM crops are imposed for protectionist reasons. It would have to make the disingenuous claim that its barriers are due to health concerns. In so doing, of course, China lends its voice to the chorus of those insisting that GM foods are, or very well might be, dangerous.

China's behavior might be explained in a variety of ways. Maybe it thinks GM crops, though useful to Chinese farmers, are also potentially dangerous and thus wants to limit the country's exposure to varieties developed and tested in Chinese labs, those best adapted to the Chinese climate, landscape, and background of wild plants and other crops. Maybe China expects to need all the food it produces to feed its own people and thus, with no export market to worry about, is not concerned about what other countries think about China's food safety.

Or maybe China is happy to add to the worldwide suspicion of GM foods because that means America will suffer that much more for its increasing reliance on them and increasing insistence that other nations buy them from us. No matter how Monsanto, the USDA, and the U.S. Trade Representative spin it, American farmers are suffering for their misplaced faith in Frankenstein foods.

Europe is the most important export market for the United States because it has the most money to spend. Europe is also the place that adheres most ferociously to the precautionary principle, in part because it has the most money to spend. It can afford to be choosy about what it will buy. Absent any destruction of their sovereignty by U.S. trade wars, European nations are likely to stick to their objections and reserve their checkbooks for food that meets their specifications. Even if the European Union does not formally ban such products forever, strong momentum already exists in the private sector to be GM-free. Big-time British retailers such as Marks and Spencer and the Anglo/French supermarket chain Carrefour loudly trumpet that their shelves are devoid of GM products, and that includes having no meat that was fed on GM corn or soybeans.

To add insult to injury, the BBC reported in 1999 that the outside catering firm that ran the employees cafeteria—"canteen" in British—at Monsanto's UK headquarters had banned GM foods from the menu. Monsanto was quick to point out that the canteen operator, Granada Food Services, had taken that step at all its operations. Still, that scarcely softened the sting of another piece of evidence that the British just will not eat GM food.

Some American farmers and farm groups cheer on their government's drive to force GM foods into world markets, but others have realized that their devotion to this technology is only making their problems worse. The American Corn Growers Association (ACGA)—the smaller, free-thinking rival of the more GM-friendly National Corn Growers Association—called in 2002 for a government study into how much U.S. agriculture's growing dependence on GM crops was costing farmers in foreign markets. The ACGA cited USDA figures showing not only that were corn prices and corn exports down, but also that the decline in corn prices had cost American taxpayers more than $1 billion in subsidies through programs that make up for market shortfalls. It also quoted a 2001 GAO report noting that reluctance in Europe and other parts of the world to buy GM corn, or even large shipments that might be contaminated by GM, was costing American farmers a share of the world market, as well as pushing the government to devote its staff and money to dealing with those restrictions and trying to coordinate export policies to get around, or through, the desires of other nations.

Europe's devotion to its precautionary principle, and to its desire to make its own decisions about whether GM and other manipulated foods are fit for their fields and for their tables, is strong and shows no signs of flagging. Even the United Kingdom, the nation with which the United States shares that oft-praised "special relationship," has decided the risks of accepting GM foods outweigh any potential benefits.

The government of Prime Minister Tony Blair had made supportive noises in regard to GM foods. But personages no less than Prince Charles and Sir Paul McCartney came down vocally on the other side, though neither was seen participating in the many acts of street and field protests or the many acts of outright vandalism aimed at GM test plots. By summer 2002 even the head of Monsanto had to admit the European market was basically closed to GM crops, at least until ongoing reviews and moratoriums were finished in 2005.

Europe, of course, is well fed and likes to support its own farmers in ways that not only keep them in business but also allow them to preserve a small-scale pastoral way of life—if for no other reason than that American tourists can take their picture. The long-term promise of biotech, after all, is supposed to be for those regions that have little to lose and, although GM marketers would never openly admit it, might gladly accept a risky source of food over the prospect of no food at all.

But in its 2002 report on the state of food security in the world, the U.N. Food and Agriculture Organization, a global body that exists for the purpose of figuring out who is hungry and what to do about it, dismisses the promise of GM crops in feeding the hungry. The fact that hunger is primarily caused by poverty, not supply, moves the FAO and others to note that countries that cannot afford plain old wheat and corn certainly will not be able to afford licensed, patented, and policed GM seed, much less the special approved pesticides and other inputs that go with them. GM crops, the FAO notes, may someday help agricultural production on marginal land—but only once the safety and environmental questions accompanying their use have been fully resolved. Until then, says the FAO, the challenge is economic, not scientific. A more equitable distribution of wealth, land, and capital, not a redistribution of genes and chromosomes, will feed the hungry people of the world.

Even the USDA's own Economic Research Service, while voicing support for GM technology, blames hunger on economic, rather than agricultural, factors. A widespread acceptance of GM crops would do nothing to relieve the problem of people being hungry because there is too little money, not because there is too little food. If anything, the property rights that the creators of GM crops insist upon keeping for themselves could make hunger a greater problem, by denying poor farmers access to licensed seed or requiring them to purchase only approved pesticides and fertilizers to use on them.

The push to save the Third World with GM foods very much resembles the Nestlé controversy of the 1970s and 1980s. People who went anywhere near an American university campus in those days may remember the "Boycott Nestlé" bumper stickers that were ubiquitous among college activists. The complaint against that multinational food corporation, best known in the United States for its chocolate bars and hot cocoa mix, was that it stood accused of tricking the mothers of Africa and South America into using Nestlé baby formula instead of relying on the traditional and more healthful habit of breast-feeding their babies. According to Nestlé's antagonists, the scam went like this:

Women who had just given birth were given free samples of baby formula mix, along with information claiming that this new miracle from the laboratories of rich people was better—or at least more modern—than mother's milk. After the women ran out of free samples, whether it was because they left the hospital or, for those who had never been to a hospital, the Nestlé employee dressed as a nurse had moved on to the next village, they would then be left to buy the powdered formula themselves, often mixing it with the local polluted water or adding too much water to stretch the supply over more time or around more babies. Even women who could not afford to continue with the formula mix might be pushed into buying it anyway, or try to share the supplies given to friends and relatives, because a few days or weeks without nursing would cause a mother's natural flow of milk to dry up.

Nestlé was accused of pushing millions of babies toward illness and death because formula mixed with untreated water caused often fatal bouts of diarrhea; others suffered woes associated with malnutrition because the formula was so watered down. One round of boycotts ended in the 1980s when Nestlé announced that it would abide by a World Health Organization (WHO) code for infant nutrition. But a boycott was again called for in the late 1990s, with fresh accusations from activist groups that Nestlé was again pushing formula in the Third World, sometimes distributing it in packaging with the WHO-approved warnings written in languages the locals could not understand.

Nestlé, while admitting past overzealous conduct, denies that it is currently engaged in any marketing program that deceives poverty-stricken women or earns itself the contempt of rich liberals. Even if the company has reformed since the 1980s, its history remains a clear example of the way modern science, with its belief that man-made solutions are superior to natural ones, is more likely to serve wealth than to serve the poor. Just as women in poor countries were once, and perhaps still are, shamed into believing that an expensive chemical made far away was better than the free milk that flowed from their own breasts, farmers of both genders in today's Third World are expected, and tempted, to believe that expensive, gene-spliced, chemical-dependent forms of agriculture brought in by rich folks from the North are better than practices and varieties that have been adapted to their climates and cultures over the centuries.

Of course, just as some mothers will always need supplements to raise healthy babies, even the most venerable farming methods can suffer everything from bad weather to waves of pests. Intelligent, careful adoption of pest-control methods, fertilizers, and new varieties of seed can indeed enhance the productivity of Third World agriculture. But just as a healthy mother can lose her ability to breast-feed through dependence on chemical substitutes, a healthy agriculture can find its genetic vitality robbed and its natural fertility exhausted by abandoning crop mixtures and rotations that keep pests guessing and people fed.

Even when GM food is offered as a gift, as it frequently is by the United States, not everybody thinks it is such a generous offer. In summer 2002, as regions of southern Africa were again facing the prospect of food shortages and widespread hunger, the governments of Zambia, Zimbabwe, and Mozambique rejected offers of U.S. corn because it would contain thousands of tons of the genetically modified variety.

"We will rather starve than get something toxic," Zambian President Levy Mwanawasa was quoted as saying in the Johannesburg newspaper *Business Day,* while the U.N. and the United States repeated their contentions that GM food was safe to eat—U.S. Agency for International Development head Andrew Natsios told the BBC he and his family had been eating it for seven years—the Zambian agriculture minister accused the donor nations of having lied to needy African peoples about what they were being fed.

The dispute showed up at the 2002 Earth Summit in Johannesburg. David King, a scientific adviser to the British government, denounced the U.S. habit of trying to sell and give GM food to African nations as a "massive human experiment." According to British press reports at the time, it was Ethiopia, a nation that has often suffered from hunger of the most telegenic kind, that blocked a U.S. move to set free-trade principles and WTO rules above the right of individual nations to reject the import of GM foods. *The Independent's* Geoffrey Lean reported that "Originally, the only resistance to the proposals came from Norway and Switzerland but after the Ethiopian delegation made its intervention the rest of the Third World swung against it. . . . The [United States] was left isolated."

A brief statement from Agriculture Secretary Veneman contained a full-throated defense of biotech by putting the question in stark terms. "Now is not the time to inflame the debate about biotechnology," she said. "Now is the time to feed starving people."

Of course, biotech apologists maintain that is the choice. Accept biotech or starve. Or, perhaps worse, push someone else to starve.

But that simple choice of eating the GM corn or starving was not really put to anyone in Africa that summer. Zambia announced it had been able to import some 300,000 tons of non-GM corn, more than enough to replace the 50,000 tons of modified corn the United States was offering. Zimbabwe and Mozambique adopted the Namibian solution of milling the corn before it was distributed. The fact that some nations are willing to distribute milled corn rather than whole kernels demonstrates that the real fear is not people eating GM corn but planting it.

Any hopes a nation—at least any nation without the international clout of the United States—may have of exporting its agricultural products is dependent on keeping those products free of GM taint. With an outright ban on the import of GM foods ongoing in Europe, and other nations requiring that GM foods be labeled as such, any country that even dreams of being a food exporter must worry about keeping its grain streams accurately labeled, if not utterly devoid of genetically modified products. The so-called developing nations already face disproportionate barriers to selling internationally, primarily in the form of high import tariffs and antidumping claims that are common in the industrialized world. The burden of maintaining a certifiably GM-free flow of grain, already discouraging to some American farmers, would be more than the developing nations could be expected to handle any time soon.

Imagine the uproar that would be raised by the U.S. State and Agriculture departments if genetically modified corn and soybeans were European or African inventions, and if those nations were threatening to use every trick in the book to ensure that

U.S. consumers bought those products. Imagine that those countries opposed any kind of regulation the United States might impose that required imported GM foods to be kept in a separate supply stream, or simply labeled as being made of genetically engineered plants. Imagine that the EU went to the WTO, claimed that there was insufficient scientific proof that GM plants, or milk and beef laced with growth-stimulating hormones, were harmful to human health and thus any regulation of them whatsoever would not be a reasonable health measure but an unreasonable trade barrier.

That is exactly what the U.S. government is doing in its efforts to get American GM products accepted in Europe, and anywhere else they are unwelcome, for no other reason but that we have the clout to demand it.

That, more than anything else, is what the GM debate is about. Clout. Not science or agriculture or even hunger, but clout. Like the questions doubtless being addressed in other meetings under way in winter 2002 at the Pentagon, one subway stop away from that USDA forum, the question of biotech is a question of power.

In recent years, the quest for private biotech power has played out in three other interesting cases: Once the industry lost. Once it fought to a sort of draw. Once it won, in a way that the world may come to regret.

The loss, from Monsanto's point of view, was its 2004 decision to abandon—maybe for a few years, more probably for good—development of a Roundup Ready wheat variety to join its pesticide-tolerant strains of cotton and soybeans. The research, which had been ongoing in partnership with the Canadian government's agriculture establishment, was aimed at getting onto the market a wheat seed that could be sprayed with Monsanto's

Roundup herbicide at times when other wheat types would be too fragile to withstand it. Because weeds and the crop they threaten are generally most vulnerable to a herbicide at the same time, the advantage is that Roundup Ready wheat can stand up to the chemical while the surrounding weeds die. The same principle underlies Roundup Ready soybeans, cotton, and canola in the nations that allow GM seed—primarily the United States, Canada, and Brazil.

Notwithstanding acceptance of Roundup Ready varieties by farmers in North and South America, importers, processors, and retailers in Europe, particularly Great Britain, were making it clear that they wanted nothing to do with genetically engineered wheat. If you grow it, came word from the EU, we won't buy it. And we'll be testing shipments and samples to make sure you haven't tried to slip some on us, or allowed the GM wheat to contaminate the real stuff. Remembering that traces of StarLink, a variety of GM corn unapproved for human consumption, was still showing up in corn samples years after its improper presence had been discovered, the Canadian Wheat Board was one of the influential organizations that wanted no part of the industry's new push, even a little bit.

Score one for the customer being right.

The split decision, which tilted mostly in favor of Monsanto, also involved Canada and the Roundup Ready line of seeds. The Canadian division of the company had sued a seventy-year-old Saskatchewan farmer named Percy Schmeiser when its roving seed snoops determined that he had been growing Monsanto's Roundup Ready canola, a crop grown for its oil, without having properly bought the seed from Monsanto, paid the fees, or signed the required agreements and licenses restricting its use to one year's planting. Schmeiser, who lost the suit in the lower courts and made his matter a cause célèbre all the way up to the Canadian Supreme Court, maintained that the crop was grown strictly from seeds he had saved from his previous year's crop. If there was Monsanto intellectual property in those seeds, he insisted, it could only be because they migrated from someone else's field.

The highest court in Canada split the difference. It ruled, five-four, that Schmeiser had indeed violated the patent on Roundup Ready canola. But it also ruled that he did not have to pay the huge judgments or fees Monsanto had been seeking because although he had planted and reaped the crop, he had not reaped any of the special benefits of using the newfangled plant. He never sprayed his field with Roundup, thus the special Roundup Ready qualities, the part of the plant that made it worth patenting, never came into play. Given the more severe lower-court rulings, Schmeiser had to count that as a legal victory, at least for his bank account. But Schmeiser and his followers remain opposed to the very idea that a farmer cannot plant whatever he finds around him without getting permission, and a receipt, from a far-away corporation that thinks a human enterprise can own a form of plant life.

The win for the biotech industry, if only because it may be an unspeakable loss for everybody else, is the previously mentioned news that corn being grown in parts of Mexico, the birthplace of the domesticated plant that feeds us all, now as likely as not contains traces of the American GM corn supposedly not allowed in that country. That is a serious matter, not only because a flood of cheap corn, GM and otherwise, has followed the North American Free Trade Agreement (NAFTA) into Mexico, depressing the market and dislodging as many as 1.3 million small farmers from their operations, but also because the introduction of GM corn threatens to outpropagate and displace the hundreds of different varieties of corn that were born and still live there.

Losing those genetic lines would threaten the entire world's corn crop, not just Mexico's, because most of the corn grown in places other than Mexico is of only a few strains that, like any monocultured crop, are susceptible to being decimated by an unknown pathogen. Keeping a rudimentary store of corn varieties, as Mexican farmers have uniquely done, is an irreplaceable bank account of genetic information that farmers will want to revisit someday.

For centuries, professional and amateur breeders have selectively bred new strains and forms of grain, as well as domestic animals, by the old-fashioned means of crossing two or more varieties of a life form in hopes of producing offspring with the positive traits of all its ancestors and none of the negative ones. The natural tendency of plants to put much of their growth energy into their stalks and roots and less into seeds—the part we eat—has been overcome by breeding since the dawn of agriculture. But as the energy choices of the plant are shifted from stalk to seed, more artificial nutrients are required.

Plant breeders, of course, have always faced the Pooh problem. A. A. Milne's Winnie-the-Pooh is informed by his friend Piglet that one could get a yummy (to Piglets) haycorn-producing oak tree by planting a haycorn. Pooh then speculates on the chances of getting a yummy (to Poohs) honey-producing beehive by planting a honeycomb. Pooh further wonders if he could get only the lovely honey, and not the stinging bees, by planting half a honeycomb. But even The Bear of Very Little Brain quickly realizes that if he planted the wrong half of the honeycomb, he could get the bees without the honey.

Traditional plant breeders, of course, come up with their share of bees-only honeycombs. But because they are always watching the progress of hundreds or thousands of individual hybrids grown in a great many different test plots, they also manage to create enough honey to make their work worthwhile and to continue selectively breeding ever better, tougher, and more productive crops. The original genetic line is not destroyed in creating the new one. Any time traditional breeders wander down a dead-end road to a bees-only honeycomb, they can easily retrace their steps back to the main thoroughfare and thence on to a more honey-producing destination.

Also bred into the traditional plant-breeding system is a concept the new biotech industry must see as totally outdated—sharing. Traditionally, new strains of grain are freely shared with other breeders, scientists, and growers, who rejigger the mix in that many more ways, wholly consistent with the kind of natural selection that has been occurring since the beginnings of life on earth. Varieties bred by commercial outfits are sold to farmers in mass quantities, enough to recover the developer's costs and fund creation of the next generation of hybrid corn, wheat, or soybean. Farmers, universities, even competing commercial breeders, meanwhile, are free to build upon the work of others to develop still more varieties of plants that may be better suited to a particular climate, soil type, or just personal tastes.

The comparison that has been made is to a lighted candle. If I have a candle burning, and you hold your unlit candle up to mine, your candle begins to glow while mine loses nothing. I can share my flame with the next person, while you go on to give illumination to still more candles. Not only does each successive lighting of a candle cost the previous owner nothing, but spreading the light this way also serves as insurance for the moment when my original candle may fail and I may need to seek a spark from someone I helped, or someone helped by that person, or anyone on down the line.

But the very essence of genetic modification, at least as it is practiced in the American capitalistic system, stands opposed to that grand and beneficial tradition of life as public domain. The high expense and long turnaround time involved in creating new genetically modified crops—though somewhat leavened in the United States by a regulatory process much more lax than that required for, say, a new pesticide—has pushed the companies involved in the business to focus as much on their products' patents as on their DNA.

WASTE LAND

My solitude grew more and more obese, like a pig.

—Mishima Yuko, *Temple of the Golden Pavilion*

It is hard to pinpoint the precise moment when the popular image of the American farmer shifted from Thomas Jefferson to Fred Ziffle.

A nation born in admiration of and dependence on the independent yeoman farmer—the salt of the earth, the chosen people of God—somehow shifted over two centuries from being a place where presidents grew up on plantations or were born in log cabins to become a nation of people who freely refer to farmers and other rural residents as "hicks" and "rubes." Wendell Berry, the novelist, poet, essayist, and farmer from Kentucky, found the prejudice alive in his own state, a state with a long tradition of independent farmers, in the voices of high school students taunting their rivals at a basketball game: "Go back, go back, go back to the woods. Your coach is a farmer and your team's no good."

In the late 1960s, CBS, the high-class "Tiffany network," built its prime-time schedule around shows premised on the idea that rural and small-town folks, though primarily honest and decent, were slow, dumb, backward people whose frequent culture shock was fair game for derision. The Clampetts of *The Beverly Hillbillies* and the Hooterville neighbors of expatriate New Yorkers

Oliver and Lisa Douglas on *Green Acres* were inhabitants of another planet where the locals couldn't count ("eleventy-ten, eleventy-eleven") or install a telephone. Farmer Ziffle treated his pig, Arnold Ziffle, like one of the family. Mayberry's Andy Taylor was a bright guy possessed of a homespun wisdom, but looked smart primarily because he was usually surrounded by dummies.

Whether cause or effect, the decline of the farmer's popular image has coincided with the decline of the population of farmers. It is as if the loss of a way of life for so many Americans must be rationalized by believing that no intelligent person would want to be a farmer and that progress, in this land of inevitable progress, must include the destruction of an entire class of isolated, ignorant, and deprived people—for their own good.

The decline in the number of farmers obviously did not translate to a decline in the amount of food, at least not in American supermarkets. That led to questions about the need for any sort of government farm programs, which led to the joke about the Department of Agriculture official who was seen sitting at his desk one day, crying. When his secretary asked what the matter was, he replied through his sobs, "My farmer died." (A similar joke is told about the Bureau of Indian Affairs.)

The shift of farming from a life of being as one with the earth to a career, a job, of using machines and chemicals to try to bring that world to heel has long been justified as a battle against the mind-stunting and unrelenting toil of farm life. There is still much of that kind of work to do, though increasingly it is being done by low-paid workers, who do not own the land they work on, have no roots in the communities where they live, and receive little reward for their labor. They do not know the land well enough to detect insect infestations or disease outbreaks early and deal with either in the most effective way. And they do not know the communities well enough to notice a suspicious stranger who might be there to carry out some heinous spy mission. Increasingly, the work is done by illegal aliens, which can lead to social tensions and more stress on the basic safety net.

The result of that has been a self-fulfilling prophecy of sorts, a depopulation and desolation of large swaths of the High Plains, leaving the remaining work of raising crops and livestock to fewer people, larger machines, stronger chemicals, and, for both plants and people, much shallower roots. That, in turn, makes agriculture more destructive of both the land and the human psyche. And it makes the crops being grown that much more vulnerable to disease, failure, even terrorism.

In the Great Plains, over most of the twentieth century, there was less a panic than a growing despair, with the realization growing rapidly in the 1980s and 1990s that agriculture was on the wrong track. The Center for Rural Affairs, with offices in Walthill, Nebraska, has been chronicling the issues involved, and the increased concern can be seen in the titles of its reports. In 1989 it was the academically straightforward "A Socioeconomic and Demographic Profile of the Middle Border." In 1990 came the less academic, cautionary "Half a Glass of Water." In 2000, as the depth of the problem was setting in, was "Trampled Dreams: The Neglected Economy of the Rural Great Plains." In 2003 deeper woe, leavened with hope, appeared in "Swept Away: Chronic Hardship and Fresh Promise on the Rural Great Plains."

The 2003 report documents how population and economic activity of the Great Plains—Kansas, Nebraska, Iowa, Minnesota, South Dakota, and North Dakota—have stagnated or declined. Of the 503 counties in those six states, 182 earn at least 20 percent of their income from agriculture. Between the census of 1990 and the count of 2000, those ag-based counties lost 9 percent of their population, even as the six-state region as a whole grew in population by 7 percent, and the officially designated metropolitan areas within the region grew by 12 percent. Thirteen percent of the households in the rural counties lived at or below the poverty level; only 8 percent of the people in the urbanized counties did.

The longer view, as James R. Dickenson took in *Home on the Range: A Century on the High Plains,* shows that many towns and counties in this region that produces much of the world's corn, wheat, and beef have declined in population by 50 percent since 1930. A good 10 percent of that decline was the result of a 1956–1957 drought and another 10 percent was due to the farm crisis of the 1980s. Nebraska alone, Dickenson reports, has an estimated 5,000 to 10,000 abandoned farmhouses.

Businesspeople fail all the time, of course. In Silicon Valley, one, two, or even three bankruptcies can be a badge of honor, a sign that someone tried something innovative, failed, and started anew. A failed dot-comer may have to give up the office, the BMW, the fancy loft, perhaps even his or her spouse, but such a failure leaves the entrepreneur to walk away with what really matters: his or her brains—plus a Rolodex of useful contacts— and thus a chance to succeed in the next venture.

When farmers go broke, they lose the basis of their existence, the hope for their future. They lose their land. In too many cases, they surrender every shred of human pride they had left. "When you're the fifth generation to work the farm and you don't succeed, you feel like a failure to your family and your community," Kansas Farmers Union President Donn Teske told *Insight* magazine. "After all, four other generations made it work and you didn't."

Those who don't totally fail often must depend on the kindness of strangers, or of the government, in ways that also do violence to the traditional rural values of self-reliance.

Poverty, as measured by the federal government, is an increasingly rural phenomenon. Many of the poorest counties in America, including eleven of the bottom twenty, are not in the Rust Belt or even the Old South, but in Nebraska and South Dakota. The 1999 per capita income of Loup County, Nebraska, was only $4,896, so small as to be dwarfed even by the nineteenth poorest, Slope County, North Dakota, where per capita income that year was $12,097.

The number of farmers who have to feed their families with food stamps or from local food pantries run by churches or other charitable organizations is more appalling than it is ironic, and this squeeze illustrates how modern farming is less about land and hard work than it is about cash flow.

There have always been pockets of rural poverty in America, of course, but the popular image was that they existed not in the farming heartland but in coal country, among miners and their families who labored in dirty and dangerous conditions to extract coal, iron ore, or other useful rocks that would leave the area but return little in the way of wealth. City-dwellers of all social classes received the fruits of the miners' labor, along with the depersonalization and pollution driven by a coal-and-iron economy. The miners, because they did not own the land or the minerals beneath it, were dependent on the whims of the mining companies for their livelihoods. Whenever a certain mine was tapped out or yielded too little minerals for the money invested, it would be closed, and the services of the miners would no longer be required. The moneymakers were the mine owners and the railroads that hauled their extractions.

Note that it is more accurate to call mining "mineral extraction," rather than "mineral production"—and that includes oil—because even the most hardworking human beings do not produce minerals. The earth produces them. Human beings only dig them out and turn them into something else, in many cases into energy that can only be used once, leaving behind large amounts of harmful waste.

Today's American farmer has much in common with his coal-mining brother, and the similarities grow with every passing year. Like nineteenth-century coal miners, twenty-first-century farmers are dependent upon others—the banker, the government, the fertilizer supplier, the commodities broker, as well as the elements themselves—for their living. Whether the food that results makes for healthful eating is no more a concern of the agribusiness system today than whether coal made for healthful breathing mattered to the coal-and-iron industry a century ago.

In most academic, government, and business circles, the word "producer" has replaced the term "farmer." To the industrial mind, that is not only a more universal, interchangeable-part term, it also is intended as a compliment. Farmers, after all, play the role of hicks and hayseeds in the industrial mind-set. To be a producer, in that way of thinking, is more honorable and certainly more profitable. Some old farmers and friends of farmers lament the new term for their way of life, but even it is not wholly accurate: As others wiser than I have noted, today's cropland is not being so much farmed as mined.

The accumulating black hole of agricultural industrialization becomes more powerful as it grows. A system based on maximizing production lowers market prices to the point that farm families can make ends meet only with government aid and off-farm income. That creates a desperate, if not altogether eager, labor pool for the huge nonunion beef, pork, and, farther south and east, chicken confined animal feeding operations and meat-processing plants. It also creates small cities that, no longer able to survive on the commerce generated by fewer, poorer farmers, are ready, willing, and able to become host to such protein factories. Once centered in urban areas—and union strongholds such as Chicago, St. Louis, and Kansas City, these protein factories are increasingly found in smaller, nonunion communities such as Holcomb, Kansas; Greeley, Colorado; Lexington, Nebraska; Logansport, Indiana; Vienna, Georgia; and Guymon, Oklahoma.

These plants have such a high demand for workers—and such a high turnover due to low pay, physically crippling work, and nauseating odors—that once-small towns now attract waves of newcomers. The new folks are often Mexican and Asian immigrants who, regardless of how hard they work and how thrifty they are with their earnings, are blamed for the overcrowding of local schools, the upward pressure on rents (never totally absorbed by the trailer courts that pop up on the edge of town), and strain on police and other public services. Culture clashes and language barriers have raised their heads, but those who voice concerns about what a new plant will do to the quality of life in a relatively small city are quickly marginalized as not only antiprogress and antibusiness but racist to boot.

Human workers are chewed up and spit out almost as quickly as the animals. Because of high injury rates—everything from crushed limbs to severe carpal-tunnel crippling—animal processors disgorge a steady stream of onetime workers whose only future is taxpayer-supported disability.

Once the plant is up and running, the smell can be stupefying. The odor is not just annoying, but also is full of toxic substances such as hydrogen sulfide that can cause a variety of illnesses, including brain damage. The toxic chemicals from giant feeding and processing facilities can be detected airborne as much as five miles away. On days when production schedules and weather conditions conspire to magnify the problem, children are not allowed to play outside, many elderly people cannot venture anywhere without their oxygen tanks, and farmers who have spent their entire lives walking through cow pies suddenly double over and puke.

A power plant or chemical factory that inflicted such harm on a community, especially a big city, would almost certainly be shut down, or at least forced into compliance with environmental rules. In smaller cities, in contrast, the cancer quickly becomes inoperable. The spread of such enterprises far and wide often makes each of them the primary economic engine for an entire county—the biggest employer, the biggest taxpayer, the only thing standing between a community's future as a regional trade center and one as the next Great Plains ghost town. Economic muscle, plus the enduring prejudice that any activity pertaining to food production is too important to be messed with, enables these ecodestroyers to keep operating.

In summer 2002, when the Sierra Club documented the sickening practices of America's big beef, pork, and poultry producers in a report called "The Rap Sheet on Animal Factories," the factory operators were ready with their canned answers.

"Tyson," reported *The Omaha World-Herald*, "said it was pleased to be part of a food production system that provides people throughout the world 'with the safest, most affordable supply of food in the world.'" The same newspaper article quoted ConAgra as labeling the report "a continuation of the Sierra Club's campaign to undermine America's confidence in and love for meat and poultry."

Love for meat and poultry is a matter of individual taste. But there is no reason to have confidence in a system that is based on the premise that the only way to get sufficient amounts of food is to impoverish other nations, along with great swaths of America, and force desperate people and communities to become intensive toxic wastelands in ways that eventually spill over into the rest of the environment.

The spillover takes forms other than the chemicals that foul the water, air, and bloodstreams, problems that involve not just the security of each individual life but of the entire nation. Among the problems that already existed, but came into sharper relief after terrorism became the most important political issue, is the unstoppable flow of illegal aliens across U.S. borders. They come from all over the world, 99 percent of them looking for nothing more than a chance at a better life. A great many of them are from Mexico or some other Latin American nation and find work somewhere in U.S. agribusiness. Some people work in the hot sun to pick vegetables, carry oranges, or cut sugarcane; others man the fast-moving bloody lines at giant slaughterhouses. Those jobs—involving hard work, low pay, and physical danger— do not attract many American citizens but are the part of the illegal-alien magnet that pulls those people here.

The part that pushes them here, not by coincidence, is also American agribusiness. The huge surplus of heavily subsidized corn that U.S. farmers and brokers dump onto the market, a market greased by the North American Free Trade Agreement, has flooded Mexico at prices as much as 30 percent below what the crop cost to grow. That corn is cheaper for Mexican ranchers, food processors, and retail grocers than the corn grown there in Mexico. Even with cheaper labor and other lower costs, Mexican corn cannot compete against U.S.-grown corn, even in Mexico. An estimated 1.3 million Mexican farmers and farm workers lost their jobs after NAFTA took effect in 1994, a figure that must have rippled through the Mexican economy as the people they did business with lost that many customers. Also between 1990 and 2000, the estimated number of Mexican citizens in the United States illegally climbed from 2 million to 4.8 million, from 58 percent of all undocumented aliens to 69 percent of them.

The farmers in other nations suffering from poverty and hopelessness, whether caused by U.S. agricultural trade policies or not, have tried growing crops that are not grown in the United States, such as tea, coffee, bananas, and cocoa. These farmers thus do not compete with U.S.-subsidized farmers, but they risk the ups and downs of international commodity markets. Farmers in South America grow marijuana, or the coca plant that is made into cocaine; in Afghanistan crops include the poppies that are milked to make heroin. The United States certainly has its share of marijuana cultivation, but due mostly to climate and habit, not many poppy or coca fields.

It is a sad symptom of how mechanical and synthetic U.S. agriculture has become that the illicit drug our rural communities do produce, in quantities that threaten them and everyone else, is not grown at all. It is derived from chemicals, chemicals often stolen from remote farms and put to the same use in human bodies that they were intended for when applied to corn and wheat. This artificial, yet deliberate, hyperactivity gets more corn from the land and more work from the person, even as it exhausts them both.

Methamphetamine—also known as meth, ice, crank, and "poor man's cocaine"—is one of the most widespread and deeply damaging drugs on the scene today. It is highly addictive and highly damaging to a person's health. Even people who do not use it are threatened, and not just by the possibility of being robbed by someone supporting his or her habit or struck by a driver on a stratospheric buzz. Cooking the stuff creates toxic, sometimes explosive by-products that are a threat to passersby and everyone in the building, which often includes the children of methheads. Cleaning up a house or hotel room that was used to make methamphetamine involves calling out the hazmat unit and donning the protective moon suits. Meth is nasty stuff.

Unlike such drugs as marijuana and alcohol, which are generally depressants that people use to wind down and relax, the point of meth is to get a person into superhigh gear. Unlike cocaine, which is generally much more expensive but provides a much shorter burst, the impact of one cheap dose of meth can last more than twelve hours. Thus, rather than being used for a recreational lift, meth has become a workplace drug that people resort to and then become dependent on to keep up with demanding jobs and still maintain hectic personal lives. In cities it is commonly abused by overstressed lawyers and other professionals who must stay mentally sharp for long periods. In poorer rural areas, it has been particularly rough on women, especially low-income single mothers, who must care for their children, hold down a job or two, and still fulfill a desire to go out dancing and dating. Meth is thought to keep young and not-so-young women fit, trim, and attractive.

The energy and attractiveness do not last long, however. After developing hollow eyes and an emaciated body, meth addicts can easily fall into physical or mental breakdown, then become subject to arrest, prison, and loss of career, children, and everything else the user was working triple shifts to try to gain and retain.

The deepest and most ironic evil connected to meth, though, is that it is taking an almost *Twilight Zone*–style vengeance against humanity for flooding the earth with the same high-powered nitrogen compound that is part of a common recipe for the drug.

The situation is reminiscent of Robin Williams's line, "Cocaine is God's way of telling you you have too much money." The scourge of meth is the earth's way of telling us we have too much energy bottled up and we are scattering it around to no good end.

One recipe for making meth is known as the "Nazi method," because that particular variety of the stimulant was issued to German soldiers during World War II. Its connection to Germany goes back further, though, to the pre–World War I Haber-Bosch method of extracting nitrogen from the atmosphere to make anhydrous ammonia. Its connection to rural America is that in addition to use of the stuff to blanket farmland with fertilizer, creating a surplus of nutrient energy that fouls our rivers, groundwater, and coastal waters, it is used to cook meth in remote buildings and lean-tos, creating a surplus of chemical energy that fouls our minds, bodies, and criminal justice system.

PHARAOH'S DREAM

**The best fertilizer is the footsteps
of the landowner.**

—Confucius

Members of Congress have been heard to say that the first government farm program came when Joseph had a dream. They say that, even though in the Bible story it was Pharaoh who had the dream, the one about the seven lean cattle swallowing the seven fat cattle. It was Joseph, he of the coat of many colors, who interpreted the dream to be a warning of lean times to come, and he who perceived the need to build reserves of food during the current period of plenty. His insight led him to high office and great reward.

No matter who had the dream, U.S. farm policy in the twentieth century was concerned less with storage than with production, less with saving the surplus for times of want than with flattening out the highs and lows with subsidy programs that encouraged maximum production at all times. The popular image of federal farm policy often focuses on efforts to take land out of production as a way to reduce supplies, boost prices, and keep hard-pressed farmers in the none-too-profitable business of feeding the rest of us. Such perceptions often led to jokes and serious concerns about "paying farmers not to grow," which seemed particularly pointless and cruel when everyone knew there were millions of hungry people in the world.

There are still millions of hungry people in the world. And there are many millions more whose lives are threatened by the high-intensity form of agriculture that shifts costs away from supermarket consumers and onto taxpayers—who, we are supposed to forget, are the same individuals.

When the 1996 farm bill—Freedom to Farm—abolished land set-asides and other devices intended to reduce production, production increased, farmers did not significantly shift into new crops, and demand, especially from Asian markets, temporarily tanked. The somber times were enough to make many farmers and observers, including then-Secretary of Agriculture Dan Glickman, nostalgic for the day when the USDA could order land removed from the production of crops of which there was already a glut. Controlling the supply of a valuable commodity, in order to make it more valuable, is a formula commonly followed in the private sector, with such products as oil and diamonds. But it is also employed by government, most notably when the U.S. Federal Reserve Bank, or the central bank of any other nation, raises or lowers interest rates.

The idea that a central agriculture authority could wield the same power over food commodity markets as Federal Reserve Chairman Alan Greenspan had with money markets is a reach. Financial markets react to Fed actions overnight or sooner. Farm commodity futures and contracts are speculative instruments by which people basically bet today what the price of corn, wheat, oats, or pork bellies will be six months from now, and trading in these instruments also is volatile. Prices respond to financial information, international and national news, and even weather reports as quickly as people can shout across the Chicago Board of Trade pit. But the feedback loop to the field and feedlot, pressure that would cause farmers, big or small, to grow more of this crop or less of that one, takes much longer to go full circle. Thus, a farm czar, a Greenspan of grain, would be unlikely even if we wanted one.

An abandonment of farm programs, specifically farm supports, would be just as unlikely and unwise. If the government ever eliminated the federal subsidy and disaster payments that so many farmers have come to depend on, directly or otherwise, the resulting financial disaster would ripple across the economy and across the world. Today's near-certainty of federal subsidies ensures that farmers will have some income, even if the market is not the source. That income not only pays the farmer's bills, but also makes him a good credit risk for this year's preplanting production loan and/or the frequent refinancing of the mortgage on his land and loans on his heavy equipment.

If that income stream were to dry up, the value of being a farmer would disappear along with the value of his land. Without that support, millions of farmers suddenly would be unable to pay the mortgage on their land, and no neighboring farmers would be in a position to buy it. The value of farmland would plummet, taking with it a great many banks, equipment dealers, farm supply businesses, and Main Street retailers, doctors, lawyers, and accountants. The crisis would make the 1980s farm crisis look like a burp and have a negative impact throughout the economy that would threaten national economic collapse.

Of course, that script will never play out. Farmers are not politically powerful enough to stop it, but all the people they do business with are, especially the multinational processors who are happy to buy, cheaply, the farmer's products without assuming the farmer's risks. The solution, then, from a public policy standpoint, is to continue using taxpayer money to support farmers.

No, correct that: The solution, from a public policy standpoint, is to start using taxpayer money to support farms—the land, not the person—while amending other laws and practices that harm both the land and the landholder.

The basic theory of land set-asides should be refined and emboldened so that the money is targeted at actions that protect the land, that make it possible for a farmer to adopt environmentally friendly practices that may, indeed, lower the amount of raw commodities he is able to produce but allows him to survive, even thrive, financially with the federal payments that reward him for using the land as wisely and lightly as possible. Particular payments for retiring even highly productive land—*especially* highly productive land—should join programs that encourage, with real money, efforts to keep plows and cows away from stream and river beds, thus keeping their "fertilizer" out of the water supply.

To the farmer and the taxpayer, such an approach would seem less like welfare and more like a rental agreement between a willing buyer and a willing seller. Some state and local governments, along with private initiatives such as the American Farmland Trust, are already buying development rights and conservation easements on working farmland nationwide. Under such deals, the farmer gets—that is, earns—money to return portions of his land to grass, trees, swamp, or whatever its prefarming state was. Such plots and strips should be along waterways as the best means of ensuring that chemicals or wastes found on working farms do not find their way into anyone else's life.

Alternatively, the farmer could continue farming the designated portions, but with limitations as to crops grown or methods followed. New York City's Watershed Management Program provides grants for farms located upriver of the city's municipal water supply to apply sustainable and minimally polluting practices to their properties, most small dairy and livestock operations. These are small farms that would otherwise face economic pressure either to become giant animal containment operations or to sell out to developers; either would require heavy water usage and would pollute what was left to get downriver to New York City. The $40 million the city provided for the plan not only kept some longtime family businesses operating, but also saved the city the cost of an $8 billion water treatment plant that would have been necessary if they had to clean the water instead of preventing its pollution in the first place.

Sometimes the farmer will continue to farm the leased or half-purchased land as he wishes, even to the point of using great amounts of chemicals and water. The buyers of the easement will be just as happy because it means the farmer can afford to keep the land and not sell out to a developer who wants to turn the farm into a housing subdivision or office park. That is acceptable, because the United States is losing farmland to urban sprawl at the rate of 2 acres per minute—an area equal to the state of Delaware every year. Any action that can maintain it as farmland, even relatively dirty farmland, is likely to be a better solution than to have another all-consuming, all-polluting, water-repelling, commuting-requiring tract-housing project.

The key to all federal farm programs, as well as farm preservation efforts undertaken by private groups, is the need to encourage wise stewardship and fair play and to cover farmers who want to try farming sustainably, and especially those who do the best job of it.

Sustainably, and with great variety.

Farms that produce acre upon acre of the same crop are vulnerable not only to market declines in the value of that crop, but also to whatever bug, blight, or rust that might attack it. Farmers attuned to Nature know better than to bet the farm on the success of one species, to engage in the kind of agriculture that agronomists call a "monoculture." A vast field of one plant, which grows to the same height at the same time and produces seeds of the same size, is easy for people to handle, especially if the model is a few people operating a few very large machines. It is also a prime target for the microbes, fungi, grasshoppers, and birds who happen to love that particular plant.

Imagine that you are a discerning, well-cultured, and intelligent person. Imagine that you really like chocolate. But I repeat myself. Suppose you are in a hallway between two dining rooms. One is filled only with chocolate. The other also contains chocolate, maybe a lot of chocolate, maybe more chocolate than is in the chocolate-only room. But suppose that scattered among the Hershey bars and Oreos, the Russell Stovers and the death-by-chocolate chocolate cake, there are other items in the second room. Maybe celery stalks and carrot sticks. Or maybe rocks and sawdust. Or dead squirrels and dog droppings. Which room are you going to plunder? Which room would you hide your chocolate in with the expectation that nobody would take it?

The room that is chocolate plus other items is a version of what is called "intercropping" or "integrated pest management" (IPM)—what used to be called smart farming.

Further suppose that the first time you see a room, it's full of nothing but chocolate, but the second time you see it, it's full of liver. Will you make much effort to return and, as the song about the boll weevil says, bring all your family there? Of course not.

That is the principle of crop rotation. The farmers around

Enterprise, Alabama, figured it out around the beginning of the twentieth century. Already discouraged by the declining fortunes of cotton, a crop that failed as often as not during many years of uncertain rainfall, farmers saw entire fields destroyed by an infestation of boll weevils. So they—the farmers, that is—gave up on cotton as their only crop and started planting other things, such as corn (of course) and, mostly, peanuts. Thanks to George Washington Carver's experiments, peanuts turned out to be the big cash crop, and they grew extremely well. Realizing this change might not have been made without the final fall of King Cotton, the town of Enterprise in 1919 erected a statue honoring the boll weevil. It is a large Roman goddess–style woman with her arms held high above her head, holding a large boll weevil aloft. The statue in the town square was vandalized many times over the years, obviously by people with no apparent sense of history or irony, and now resides in a place of honor in the local Depot Museum.

Despite the major shift accomplished by the farmers of Enterprise, most of American agriculture is more like the room of all chocolate, all the time. The threat of an insect, a microbe, or even a terrorist destroying a whole region's or nation's crop is slim. Still, that threat is far greater than it would be in a properly diverse system of agriculture where farms mixed row crops with grasses, livestock with grains, fish with rice paddies. In the unnatural monoculture, only a couple of deliberate infestations of fields or storage facilities with a naturally occurring plant pathogen would be needed to rattle the commodities markets and make the entire economy susceptible to blackmail and panic.

The way to approach all this politically, then, is to make it a national-security issue—which, by coincidence, it is.

Certainly, America has a history of using the National Security

Excuse to solve other problems. The Interstate Highway System was inspired by Dwight Eisenhower's perception of a need for a coast-to-coast road system capable of moving large numbers of troops and military machines. That it also revolutionized travel and tourism, created the fast-food and motel industries we know today, made truck transportation of goods cheaper—and contributed greatly to the destruction of passenger rail service and the atomization of the American family—were all side effects.

Federal funding for education took off in the wake of Sputnik and the fear that Soviet children were somehow ahead of American students in their knowledge of math and science. The Internet was created as a decentralized way for government scientists to communicate and for military operations to continue to function in case a nuclear attack "decapitated" the chain of command. The urgency and the funds for the race to the moon issued from the perceived need to beat the Soviets to the literal high ground of space.

Unfortunately, although space, highways, and even education could be approached as industrial problems manageable with industrial solutions, the problems of the food supply are caused by industrial thinking. The decentralization and creative nature of the Internet may provide some models for the proper directions for agriculture, but the problems of food will not be solved with industrial solutions because food, no matter how hard we try to rationalize otherwise, is not an industry. Food is life, which flows from Nature, which will not conform to human cleverness. For our technology to feed us, it must respect and even mimic the processes of Nature. It must not try to overwhelm Nature because, in the end, the forces of Nature far outnumber, and are far more experienced, than anything the industrial mind can throw at them.

So far, though, few individuals or groups involved in the discussion appear to see the industrial mind-set itself as the real problem. Even consumer advocates such as the Ralph Nader–founded Public Citizen or the Union of Concerned Scientists react to reports of contaminated food and toothless regulators by calling for changes that are only cosmetic. Consumer watchdogs regularly insist on enhanced inspections and stronger laws that would allow federal inspectors more power to halt production at plants that do not meet safety standards and to order the recall of suspect meat without having to wait days for lab reports to come back and certain links to be established. They rightly decry the political influence of large corporations that testify, lobby, and even sue to keep their operations rolling ahead with as little federal interference as possible.

In that approach, though, the public watchdogs sometimes seem to differ little from the official apologists for big agriculture, who see the existing system simply as a fortress that needs to be more closely guarded. Rather than the suggestion that this system needs to be uprooted, the emphasis becomes one of circling the wagons around the slaughterhouses and grain elevators, putting more money into disease-prevention research, and creating criminal penalties for so much as photographing livestock operations with any kind of malicious intent. (The last is a club that will more likely be brandished at antibiotech activists or animal-rights demonstrators than used against foreign terrorists.)

Farming, then, cannot be truly industrial, but there are business models that will serve, and serve well. After the attacks on the World Trade Center and the Pentagon, Americans rushed to install stronger doors on airline cockpits and stand in longer lines at airport security checkpoints, guarding against an unlikely, but no longer unthinkable, repeat of the horrific use of aircraft as missiles. Meanwhile, more quietly, the titans of industry that had been most affected by the attacks, the financial houses of Lower Manhattan, drew one important lesson from September 11. They began breaking themselves up, not by ownership but by location.

The New York Times reported on January 29, 2002, that Morgan Stanley, the largest securities company in Manhattan, was moving to New York state's Westchester County. Goldman Sachs and Company was contemplating the move of its equity trading department to New Jersey, and Marsh and McLennan, the insurance and financial services company left homeless by the destruction of the World Trade Center, pondered a move to Hoboken, New Jersey.

"Never again," the *Times* reported, "do these global companies want to see themselves knocked out of business for days and weeks, or even hours, by a single cataclysmic event.

"Dispersal is in."

Dispersal is something Nature has always known. It is something for those who produce our food to learn again, and for those who eat it to encourage and reward.

Afterword:
Wasps and Finches

It may seem strange that any men should dare to
ask a just God's assistance in wringing their bread
from the sweat of other men's faces.

—Abraham Lincoln, second inaugural, 1865

Stephen Jay Gould, the late scientist who was probably the most
well-known and articulate voice of evolutionary biology, described the state of the modern American farmer, even though he
probably had no intention of doing so, in his 1999 book *Rocks of
Ages*. Explaining how Charles Darwin and other naturalists of the
nineteenth century could accept the idea of evolution without
turning their backs on God—despite the charges leveled by some
religious leaders then and now that they were doing exactly
that—Gould told the story of the ichneumonid wasp.

That sort of wasp, like some other insects, has a habit of laying
its eggs on or in the body of another insect that the mother wasp
has paralyzed—but not killed—with a sting. When the young
hatch, the larvae proceed to eat the hapless cricket or caterpillar,
which is an excellent source of food because, though utterly im-
mobile, it remains alive. The larvae are even somehow clever
enough to eat the host's heart and other vital organs last, as
Gould says, "lest the host decay and spoil the bounty." This

apparent infliction of unimaginable suffering upon one living creature by another, the early Darwinians held, was a good reason to believe that animal behavior had evolved independent of a loving God. The idea that Nature happened, but was not designed, excuses a world where, in Darwin's words, "a cat should play with mice."

But Nature does not justify a world in which a shrinking but still significant, and always crucial, segment of the economy is cruelly maintained, like a caterpillar full of wasp larvae, for the benefit of others. As scientists later determined, the captive caterpillars lack both the nervous systems and the higher brain functions that would subject them to the physical pain and emotional torment suffered by, say, the cocooned humans victimized by the otherworldly monsters in the movie *Alien* and its sequels. But farmers suffer economically and psychologically because of a system that keeps them just alive enough to feed the rest of us. And whereas the relationship of the wasp to the caterpillar has continued and can continue for millennia, the cruel captivity of the modern farmer does economic, environmental, and emotional damage to the world as a whole, damage that cannot be sustained indefinitely.

The wasp's sting felt by the modern farmer, the kind that immobilizes but does not kill, at least not immediately, is the never-ending spiral of overproduction. It does not matter whether commodity prices are up or down, weather is good or bad, government is in a giving mood or cutting back, international trade is flowing or stifled. Just about everything that happens to an American farmer is seen as an incentive to increase production.

The Agriculture Department, ever the Pollyanna of farming, described the situation thus: "Toward the end of the 20th century farmers and ranchers were increasingly caught in a cost-price squeeze that required them to become savvy marketers as well as producers of agricultural commodities." Although there are many bright spots in American agriculture that do, indeed, involve smart and innovative marketing, the overarching thrust of government programs, from subsidies to research, is still geared toward encouraging and rewarding increased productivity. Playing that game well can help farmers earn more subsidies and borrow more money against the inflated value of their land so they can buy more equipment and chemicals to work the land they bought when their less successful neighbors sold out or died. Like the heart of the captured caterpillar, they will be eaten last by the worms.

There are basically two ways to maintain prices for raw or barely processed food at a living level for farmers and ranchers and yet keep consumer prices reasonable. One is to have more middlemen—processors, shippers, marketers, retailers—competing to offer the most to farmers, to win their grain, beef, or whatever, and then turn around and compete to offer the finished product to consumers for less, to win their business. The other way is to get rid of as many middlemen as possible, getting food directly from farmers to consumers so that whatever amount of money does change hands need not be shared with others.

Adding middlemen will require aggressive and consistent action by the government, both state and federal, enforcing antitrust laws, pollution laws, pure food and drug laws, and other statutes and rules that, when ignored, allow big food companies to live the lie of speed and conglomeration being necessary for survival—ours, supposedly, not theirs. Government action is possible but not likely.

Yet deleting the middlemen can be accomplished without permission of either the government or big agriculture. It requires only a decision by enough farmers and enough consumers that it is the way to go.

In his book *Lords of the Harvest,* Daniel Charles relates a story told by an entrepreneur, David Padwa, who in 1981 was trying to raise capital for a new company that would get in on the ground floor of what promised to be the boom industry of genetically modified crops. Among those he approached was George Soros, the international financier and, later, free-market philosopher. Soros wasn't buying.

"I don't like businesses where you only get to sell your product once a year," Soros explained. "And I don't like businesses in which anything you could possibly do will be overwhelmed by the effects of the weather."

Of course, Cargill, Archer Daniels Midland, Tyson, ConAgra, and the other buyers and marketers of grain, cows, pigs, and chickens are, like Soros, too smart to get into the very business he describes. The agribusiness giants do not farm the land. They farm the farmers. The farmers, in turn, run the risk of crops being ruined by flood or drought. Then, although people eat every day, the farmers generally come to market with the same crop at the same time so that no one of them has a shred of market power.

The farmers go into debt to build hog or chicken houses, praying they won't be dropped on a whim by the one corporation that is in a position to buy their fattened animals, or left holding the bag when consumer demand or environmental regulations change. The farmers make enemies of their growing number of suburban and ex-urban neighbors by spraying pesticides or spilling manure into the creek. The farmers fight with their urban neighbors for dwindling supplies of water.

Many of those who hang on manage to do so only because of

income that comes from activities other than selling their grain or their livestock on the open market, or however it is sold. An increasing number of the small farms, depending on the crops they raise, receive a relative pittance of federal support or none at all. At first glance, they might seem to be the sort of independent farm that once covered the landscape and warmed the hearts of Jefferson and Lincoln. But so many of those farms produce ever-smaller proportions of the nation's food, and what is produced not only fails to support the family but often loses money. Thus, the farm family's lifestyle is supported by one or more members of the family working in town. These farmers moonlight and commute, sacrificing the very independence and family together-ness that drew them to farming in the first place, for the benefit of other people's profitable agribusiness and the average con-sumer's low grocery bill.

Some of the lucky or clever ones manage to moonlight in ways still related to farming, selling equipment or seed to other farm-ers, even working as advisers or activists in programs that help their fellow stewards of the soil hold on to that land and make it work. Through their efforts and those of many others, there is increasing hope for more independent farms and independent styles of farming.

They are there. We have only to look, and take a little time do-ing so. Here are some examples:

- The Missouri woman who was made so sick by the foods she ate that she became virtually disabled. She started her own organic meat business, providing everything from rabbit and lamb to beef and eggs, all without stimulants, hor-mones, or even antibiotics, preferring to treat any sick live-stock with natural remedies.

- The former industrial-agriculture feed dealer in Kentucky who now raises cattle without commercial feed. His animals eat only grass, virtually eliminating the possibility that they will carry either *E. coli* poisoning or mad cow disease, in addition to providing meat that is better for the human heart.
- The taco stand in Waterloo, Iowa, that seeks to buy local foodstuffs for use in the business and in 2001 succeeded 71 percent of the time, including all of its beef, chicken, pork, and cheese.
- The Albany, New York–area organic vegetable grower who supplies 800 paying shareholders with as many as fifty different types of produce for twenty-five weeks each season.
- The self-described "disgruntled former cattle farmer," another Missourian, who supports his family by raising a complementary mix of organic vegetables, sausage, corn, and fish.
- The South Carolina dairy farmer who figured out how to make more money milking, feeding, and caring for fewer cows. The fewer cows still provide enough manure to fertilize the pastures that provide food for the contented cows, and he no longer has to buy commercial feed, truck away manure, or service his operation's large debt, which is now gone.
- The Nebraska farmer and rancher who raises pigs in a pasture rather than a confinement barn, and cows on rotating grazing rather than a smelly feedlot; he markets all of it through a local co-op, of which he is a member. Now he holds seminars advising other young people how to become his kind of farmer.

There are thousands more such operations, available to explore through the USDA's Sustainable Agriculture Research and Education Program, www.sare.org, or through the Clearinghouse for Information About Pasture-Based Animals, www.eatwild.com.

You will need to find a local food provider near you, but that is the point. Food should be intensely local. It belongs to a place. Most food worthy of being talked about is referred to by place of origin—Italian, Chinese, Thai, and so on. People who raise food well will be tied to their particular market, raising food that could only come from that place and time.

Thus, to succeed, farmers must determine what foodstuffs are settled, by climate or by habit, in their area, and adapt their operation to produce something no one else would create, some food that is perfect both for the land that will grow it and the people who will eat it. A lesson comes from the birds of that Pacific island that the naturalist on *HMS Beagle* believed existed to teach us about natural selection: The characteristics of small wild birds varied according to what food was available.

The world will be fed, if it is fed, by many thousands of different sources, each adapted and adapting to local conditions, from California to Côte d'Ivoire, as they exist and as they are modified by both internal and external forces. It will be fed not by Darwin's grim wasps but by his multiplicity of finches.

Acknowledgments

This is not a scholarly work, so the following is less in the nature of academic footnotes than it is for the purpose of giving credit where it is due. A great deal of credit is due many people, some for providing facts, others context, history, or inspiration.

The list begins and ends with my family. First my parents and siblings, who were unfailingly supportive and created an atmosphere of curiosity, good humor, and love of language. Then my wife, Rebecca, and our sons, Graham and MacGeorge, who tolerated the long periods of absence or absentmindedness while I was working on this project and provided just enough of a push for me finally to get it done.

Next, my professional associates, who also had to put up with my mind as well as my body often being elsewhere. At the Land Institute in Kansas, those people included Wes Jackson, Ken Warren, Stan Cox, Joan Jackson, Marty Bender, Lee DeHaan, Jerry Glover, David Van Tassel, Scott Bontz, Elizabeth Granberg, Stephanie Hutchinson, Darlene Wolf, Patty Melander, Bob Pinkall, Steve Renich, Chris Picone, Helen Ridder, Conn Nugent, and Harris Rayl. At *The Tribune* in Salt Lake City, Vern Anderson, Paul Wetzel, Marilyn McKinnon, Malin Foster, and Pat Bagley.

Also, I thank my original agent, Elly Sidel, who suggested to me quite out of the blue one day that I might be able to write a book, and to her successor at John Hawkins and Associates, Moses Cardona. Much credit must be given to PublicAffairs founder Peter Osnos, who took a chance, as he often does, on an unknown who had never written a book before. And to his editors Paul Golob and, later, David Patterson, Ida May Norton, and Robert Kimzey. David knew how to get me to move along without freezing me up, in part by offering good suggestions about the title and insight about how the state of affairs on the modern farm sounds like "Sharecropping 2.0" and recommending the absence of any SpongeBob SquarePants–brand lentils. Thanks also to publicity director Gene Taft and crew for all their work.

The origins of this book are in the research, travel, editorials, and musings underwritten by the Eugene C. Pulliam Fellowship for Editorial Writing, awarded by the Sigma Delta Chi Foundation of the Society of Professional Journalists. Many thanks to the judges who granted me that honor and assistance: Paul McMasters, Jean Otto, and Forrest Landon. I hope they are pleased with the outcome.

To the degree that this book is a success intellectually, financially, or any other way, the credit is theirs. Any errors, omissions, or misinterpretations are solely those of the author.

. . .

Source Notes

Throughout this book, statistics on the U.S. farm population, number of farms, their size, production, and income are drawn primarily from the following:

Food and Agricultural Policy: Taking Stock for a New Century, U.S. Department of Agriculture (USDA), Washington, D.C., September 2001.

Trends in U.S. Agriculture, U.S. Department of Agriculture, National Agricultural Statistics Service (NASS), www.usda.gov/nass/pubs/trends.

The 2002 Census of Agriculture, USDA, NASS; and *Statistical Abstract of the United States: 2004–2005,* Section 17: Agriculture.

A brief, useful, yet dispassionate overview of the issues addressed in this book is "How Sustainable Agriculture Can Address the Environmental and Human Health Harms of Industrial Agriculture," Leo Horrigan, Robert S. Lawrence, and Polly Walter, Center for a Livable Future, Johns Hopkins Bloomberg School of Public Health, Baltimore, *Environmental Health Perspectives,* May 2002.

PROLOGUE: SEARCHING FOR ROOTS

Quotes from Agriculture Secretary Ann Veneman and Kansas cattleman Mike Callicrate are from "A Food System Gone Wrong," by Callicrate on his website: www.nobull.net.

STALIN'S REVENGE

Information on the concentration of ownership in American agriculture was drawn from these materials:

The Corporate Reapers: The Book of Agribusiness, A. V. Krebs, Essential Books, Washington, D.C., 1992.

Willard Cochrane and the American Family Farm, Richard A. Levins, University of Nebraska Press, Lincoln, 2000.

"USDA Inc.: How Agribusiness Has Hijacked Regulatory Policy at the U.S. Department of Agriculture," Philip Mattera, Corporate Research Project of Good Jobs First, Washington, D.C., presented July 23, 2004, Organization for Competitive Markets Food and Agriculture Conference, Omaha, Nebraska.

A Food and Agriculture Policy for the 21st Century, ed. Michael C. Stumo and published in 2000 by the Organization for Competitive Markets, Lincoln, Nebraska, www.competitivemarkets.com.

"Multi-National Concentrated Food Processing and Marketing Systems and the Farm Crisis," William D. Heffernan and Mary K. Hendrickson, both of the University of Missouri–Columbia, presented at the annual meeting of the American Association for the Advancement of Science Symposium, "Science and Sustainability: The Farm Crisis, How the Heck Did We Get Here," February 14–19, 2002, Boston.

"Consolidation in the Food and Agriculture System," report to the National Farmers Union, William Heffernan, Mary Hendrickson, and Robert Gronski, all of the University of Missouri-Columbia, February 5, 1999.

"Economic Concentration and Structural Change in the Food and Agriculture Sector: Trends, Consequences, and Policy Options," prepared by the Democratic staff of the Committee on Agriculture, Nutrition, and Forestry of the U.S. Senate, Tom Harkin, Iowa, ranking Democratic member, October 29, 2004.

"The Corruption of American Agriculture," Tad Williams, Americans for Democratic Action Education Fund, Washington, D.C., 2000, www.adaction.org.

"U.S. Hog Breeding Herd Structure," USDA, NASS, September 13, 2002, www.usda.gov/nass/.

"U.S. Broiler Industry Structure," USDA, NASS, November 27, 2002, www.usda.gov/nass/.

A Time to Act: Report of the USDA National Commission on Small Farms, USDA, Washington, D.C., January 1998.

"Issues," Illinois Farm Bureau website, www.ilfb.org.

FROM "MORE!" TO "TOO MUCH!"

Pat Robertson quotation from "The Turning Tide," Alf Landon Lecture Series address at Kansas State University, October 1993, www.patrobertson.com/speeches/alflandon.asp.

Information on mission, history, and number of land-grant colleges is from these items:

Land-Grant Universities and Extension into the 21st Century: Renegotiating or Abandoning a Social Contract, George R. McDowell, Iowa State University Press, Ames, 2001.

"The 105 Land-Grant Colleges and Universities," National Association of State Universities and Land-Grant Colleges, www.nasulgc.org, accessed March 7, 2005.

History of U.S. agricultural policy and programs has been drawn primarily from the following:

Problems of Plenty: The American Farmer in the Twentieth Century, R. Douglas Hurt, Ivan R. Dee, Chicago, 2002. Henry Wallace quotation also from this source, p. 47.

Willard Cochrane and the American Family Farm, Richard A. Levins, University of Nebraska Press, Lincoln, 2000.

"A Short History of Agricultural Institutions," Mary Summers, in *A Food and Agriculture Policy for the 21st Century*, ed. Michael C. Stumo, Organization for Competitive Markets, Lincoln, Neb., 2000.

Directions for Future Farm Policy: The Role of Government in Support of Production Agriculture, Commission on Twenty-first Century Agriculture, USDA, January 2001.

Comments on Freedom to Farm by Sen. Pat Roberts, personal interview, Washington, D.C., February 23, 1999.

BUT WHO WILL FEED THE WORLD?

Quotation from Jared Diamond, *Guns, Germs, and Steel: The Fates of Human Societies,* W. W. Norton, New York, 1998.

Information on the Green Revolution and hunger around the world is from the following:

"Forgotten Benefactor of Humanity," Gregg Easterbrook, *The Atlantic*, January 1997, www.theatlantic.com/issues/97jan/borlaug/borlaug.htm.

"The Green Revolution: Peace and Humanity," Norman E. Borlaug, *The Atlantic*, January 1997, www.theatlantic.com/issues/97jan/borlaug/speech.htm.

The State of Food Insecurity in the World 2003: Monitoring Progress Towards the World Food Summit and Millennium Development Goals, U.N. Food and Agriculture Organization (FAO), Rome, Italy, 2003.

Agriculture: Towards 2015/30: Technical Interim Report, FAO, Rome, Italy, 2000.

The Next Green Revolution: Essential Steps to a Healthy, Sustainable Agriculture, James E. Horne and Maura McDermott, Food Products Press, New York, 2001.

"Rigged Rules and Double Standards: Trade, Globalization, and the Fight Against Poverty," Oxfam, 2002, www.oxfam.org.uk.

"Running into the Sand: Why Failure at the Cancun Trade Talks Threatens the World's Poorest People," Oxfam briefing paper, August 2003, www.oxfam.org.uk.

"United States Dumping on World Agricultural Markets," Institute for Agriculture and Trade Policy, Minneapolis, Minnesota, 2003, www.iatp.org.

Conflicting *Farm Journal* articles: "Outlook: Surprise, Surprise," Bob Utterback, and "From the Field: The Year for Fungicides," Ken Ferrie, *Farm Journal*, November 2004.

"U.N. Chief Urges Rich States to Drop Farm Subsidies," Reuters, June 10, 2002.

"Your Farm Subsidies Are Strangling Us," Amadou Toumani Tour and Blaise Compaor, Op-Ed column, *New York Times*, July 11, 2003.

THE MONEY FAMINE

This chapter draws primarily from *Poverty and Famines: An Essay on Entitlement and Deprivation*, Amartya Sen, Oxford University Press, Oxford, 1981.

Jacques Diouf quotation from *The State of Food Insecurity in the World 2003: Monitoring Progress Towards the World Food Summit and Millennium Development Goals,* FAO, Rome, Italy, 2003.

Career and Nobel award for Amartya Sen from "Indian Wins Nobel Award in Economics," Sylvia Nasar, *New York Times,* October 15, 1998.

Hendrick A. Verfaillie quotation from "Securing Our Commitments to Agriculture," report delivered at *Farm Journal* conference, Washington, D.C., November 27, 2001.

Events in Argentina from *New York Times* articles: "Argentine Cabinet Offers to Quit as Death Toll in Riots Rises to 16," Clifford Krauss, December 20, 2001; "Argentine Food Riots End, but Hunger Doesn't," Larry Rohter, December 24, 2001.

DON'T HELP THAT BEAR

Gilles Stockton quotation from "Ranching's Worst Enemy? It's Not Greens," Ray Ring, *High Country News,* March 15, 2004.

Events involving Creekstone Farms from "Meatpacker: USDA Policy on Mad Cow to Blame for Layoffs," Dave Ranney, *Lawrence* (Kansas) *Journal-World,* January 2, 2005.

Figures on farm size and output from these sources:

"The Multiple Functions and Benefits of Small Farm Agriculture in the Context of Global Trade Negotiations," Peter M. Rosset, Food First Policy Brief no. 4, September 1999, quoting statistics from 1992 USDA Agricultural Census.

"Are Large Farms More Efficient?" Staff paper P97-2, Willis L. Peterson, Department of Applied Economics, College of Agricultural, Food, and Environmental Sciences, University of Minnesota, January 1997.

Information on U.S. farm subsidies from the following:

"Another Year at the Federal Trough: Farm Subsidies for the Rich, Famous, and Elected Jumped Again in 2002," Brian M. Riedl, Heritage Foundation Backgrounder, May 24, 2004.

Environmental Working Group Farm Subsidy Database, www.ewg. org/farm.

"The Farm Bill: A Twice-Baked Potato," Heather Lawhon, National Center for Policy Analysis, Dallas, Texas, Brief Analysis no. 413.

"Payment Paradox: Subsidy Recipients Defend Payments from the Government," Tim Unruh, *Salina (Kansas) Journal*, December 8, 2004.

Information on enforcement of U.S. antitrust laws dealing with agricultural production and retailing from these sources:

"The Structural Transformation of the Agriculture Sector," Neil E. Harl, presented at conference "Fixing the Farm Bill," National Press Club, Washington, D.C., March 27, 2001.

"Economic Concentration and Structural Change in the Food and Agriculture Sector: Trends, Consequences, and Policy Options," prepared by the Democratic staff of the Committee on Agriculture, Nutrition, and Forestry of the U.S. Senate, Tom Harkin, Iowa, ranking Democratic member, October 29, 2004.

"The Corruption of American Agriculture," Tad Williams, Americans for Democratic Action Education Fund, Washington, D.C., 2000, www.adaction.org.

Information on *Henry Lee Pickett et al., v. Tyson Fresh Meats, Inc.*, from Plaintiff's Third Amended Complaint, February 16, 2000, United States District Court for the Middle District of Alabama; and Brief of Amici Curiae Fifty Leading Scholars . . . , June 16, 2004, United States Court of Appeals for the Eleventh District; "Meatpacker Consolidation Draws Concern," Emily Gersema, Associated Press, June 19, 2002.

THE RICH GET FATTER, THE POOR GET . . . FATTER

Information on how the price of food is not "elastic" comes from the following:

Directions for Future Farm Policy: The Role of Government in Support of Production Agriculture, Commission on Twenty-first Century Agriculture, USDA, January 2001.

"Rethinking U.S. Agricultural Policy: Changing Course to Secure Farmer Livelihoods Worldwide," Daryll E. Ray, Daniel G. De La Torre Ugarte, and Kelly J. Tiller, Agricultural Policy Analysis Center, University of Tennessee, Knoxville, www.agpolicy.org/blueprint.html.

Figures on personal income spent on various foods, and farmers' share of the retail food dollar, are from Economic Research Service of the USDA, www.ers.usda.gov/briefing.

Information on obesity and malnutrition from these sources:

"Obesity: The Developing World's New Burden, Focus," FAO, January 2002, www.fao.org/FOCUS/E/obesity/obs1.htm.

Overfed and Underfed: The Global Epidemic of Malnutrition, Gary Gardner and Brian Halweil, Worldwatch Institute, Washington, D.C., March 2000.

"Halting the Obesity Epidemic: A Public Policy Approach," Marion Nestle and Michael F. Jacobson, *Public Health Reports,* January/February 2000.

"The Price Is Right," Jayachandran N. Variyam, *Amber Waves,* Economic Research Service, USDA, February 2005.

"Surgeon General Outlines National Plan on Obesity," Sally Squires, *Washington Post,* December 13, 2001.

Battle of the Bulge, University of Nebraska–Lincoln, College of Journalism and Mass Communication, 2002.

"How Can Californians Be Overweight and Hungry?" Patricia B. Crawford et al. *California Agriculture,* January–March 2004.

"Fighting Hunger Today Could Help Prevent Obesity Tomorrow," FAO, via PRNewswire, February 12, 2004.

"Is the U.S. Government's Farm Policy Making You Fat?" Alan Bjerga, *Wichita* (Kansas) *Eagle,* November 28, 2004.

"The Fat of the Land," Scott Fields, *Environmental Health Perspectives,* October 2004.

TO HELL IN A BUSHEL BASKET

Much of the information in this section is from work by University of California professor Michael Pollan, including "When a Crop Becomes King," *New York Times,* July 19, 2002; "The (Agri)Cultural Contradictions of Obesity," *New York Times Magazine,* October 12, 2003; and "A Flood of U.S. Corn Rips at Mexico," *Los Angeles Times,* April 23, 2004.

Information on the link between a corn diet for cattle and *E. coli* poisoning in humans is from "Acid Relief for O157:H7: Simple Change in Cattle Diets Could Cut *E. Coli* Infection, USDA and Cornell Scientists Report," *Cornell News*, September 10, 1998, www.news.cornell.edu/releases/Sept98/acid.relief.hrs.html.

Other sources include the following:

"Blaming It on Corn Syrup," Patricia King, special to *Los Angeles Times*, March 24, 2003.

"National Diabetes Fact Sheet," Centers for Disease Control and Prevention, www.cdc.gov/diabetes/pubs/estimates.htm#costs.

"CU Scientist Terms Corn-Based Ethanol 'Subsidized Food Burning,'" *Cornell Chronicle*, August 23, 2001, www.news.cornell.edu/Chronicles/8.23.01/Pimentel-ethanol.html.

"The Cost of Ethanol," editorial, *National Post*, June 19, 2003.

DROWNED AND DRAINED

"Nitrogen's Deadly Harvest," Heather Dewar and Tom Horton, *Baltimore Sun*, September 24–28, 2000.

America's Living Oceans: Charting a Course for Sea Change, A Report to the Nation, Pew Oceans Commission, Leon E. Panetta, chair, May 2003, www.pewtrusts.org/pdf/env_pew_oceans_final_report.pdf.

"Confined Animal Production and Manure Nutrients," Agriculture Information Bulletin no. 771, Economic Research Service, USDA, Washington, D.C., June 2001.

"The Dead Zone," *Economist*, August 24, 2002.

Dust Bowl: The Southern Plains in the 1930s, Donald Worster, Oxford University Press, Oxford, 1979.

"Aqueous Solutions," chap. 11 in *Natural Capitalism: Creating the Next Industrial Revolution*, Paul Hawken, Amory Lovins, and L. Hunter Lovins, Little, Brown, New York, 1999.

"Farming in the Public Interest," Brian Halweil, chap. 3 in *State of the World 2002*, Worldwatch Institute, W. W. Norton, New York, 2002.

WHY RACHEL CARSON STILL MATTERS

Pertinent sources for this chapter include these:

Silent Spring, Rachel Carson, with essays by Edward O. Wilson and Linda Lear, 40th anniversary ed., Houghton Mifflin, Boston, 2002.

"Food for Life," chap. 10 in *Natural Capitalism: Creating the Next Industrial Revolution,* Paul Hawken, Amory Lovins, and L. Hunter Lovins, Little, Brown, New York, 1999.

Living Downstream: An Ecologist Looks at Cancer and the Environment, Sandra Steingraber, Addison-Wesley, Reading, Mass., 1997.

Why Poison Ourselves? A Precautionary Approach to Synthetic Chemicals, Anne Platt McGinn, Worldwatch Paper no. 153, Worldwatch Institute, Washington, D.C., November 2000.

Unreasonable Risk: The Politics of Pesticides, Center for Public Integrity, Washington, D.C., 1998.

"Why Farmers Continue to Use Pesticides Despite Environmental, Health, and Sustainability Costs," Clevo Wilson and Clem Tisdell, School of Economics, University of Queensland, Brisbane, Australia, *Ecological Economics,* December 2001.

"Pesticides: Their Multigenerational Cumulative Destructive Impact on Health," Report to The House of Commons, Standing Committee on Environmental and Sustainable Development, from the Environmental Illness Society of Canada, February 2000.

"Trouble on the Farm: Growing Up with Pesticides in Agricultural Communities," Natural Resources Defense Council, n.d., www.nrdc.org/health/kids/farm/exec.asp

"Analyzing the Ignored Environmental Contaminants," Britt E. Erickson, *Environmental Science and Technology,* April 1, 2002.

"Semen Quality in Relation to Biomarkers of Pesticide Exposure," Shanna H. Swan et al., Environmental Media Services, Washington, D.C., June 2003, www.ems.org/male_reproductive/missouri_study.html.

"Global Cancer Rates Could Increase by 50% to 15 Million by 2020," World Health Organization, press release, April 3, 2003.

"Hormones: Here's the Beef: Environmental Concerns Reemerge over Steroids Given to Livestock," Janet Raloff, *Science News,* January 5, 2002.

"Monsanto: Food, Health, Hope," A. V. Krebs, *Food First Back-grounder,* Summer 1999.

"Opinion of the Scientific Committee on Veterinary Measures Relating to Public Health: Assessment of Potential Risks to Human Health from Hormone Residues in Bovine Meat and Meat Products," European Commission, Directorate-General, April 30, 1999.

"Warning: Corporate Meat and Poultry May Be Hazardous to Workers, Farmers, the Environment, and Your Health," Marc Cooper, Peter Rosset, and Julia Bryson, *Food First Backgrounder,* Spring 1997.

"Hogging It: Estimates of Antimicrobial Abuse in Livestock," Margaret Mellon, Charles Benbrook, and Karen Lutz Benbrook, Union of Concerned Scientists, Washington, D.C., January 2001, www.ucsusa. org/publications.

SOUNDS LIKE SCIENCE

Dinner at the New Gene Café: How Genetic Engineering Is Changing What We Eat, How We Live, and the Global Politics of Food, Bill Lambrecht, Thomas Dunne Books, St. Martin's Press, 2001.

Lords of the Harvest: Biotech, Big Money, and the Future of Food, Daniel Charles, Perseus Publishing, 2001.

"Who Benefits from Biotechnology?" Michael Duffy, Iowa State University, presented at the American Seed Trade Association meeting, Chicago, December 5–7, 2001.

"Impacts of Genetically Engineered Crops on Pesticide Use in the United States: The First Eight Years," Charles M. Benbrook, BioTech InfoNet Technical Paper no. 6, November 2003, www.biotech-info.net.

"Do GM Crops Mean Less Pesticide Use?" Charles Benbrook, *Pesticide Outlook,* Royal Society of Chemistry, Hemel Hempstead, U.K., October 2001.

"Don't Eat Again Until You Read This," Jeff Wheelwright, *Discover,* March 2001.

"Environmental Effects of Genetically Modified Food," Margaret Mellon and Jane Risser, Union of Concerned Scientists, Washington, D.C., June 2003, www.ucsusa.org.

"GM Crop Taints Honey Two Miles Away, Test Reveals," *Sunday Times* (London), September 15, 2002.

"Unraveling the DNA Myth," Barry Commoner, *Harper's*, February 2002.

"A Biotech Fix for Hunger?" Brian Halweil, *State of the World 2002*, Worldwatch Institute, W. W. Norton, New York, 2002, p. 58.

A TARGET-RICH ENVIRONMENT

Tommy Thompson quotation from "Thompson Steps Down," Deborah Barfield Berry, *Newsday*, December 4, 2004.

Charles Beard and Harley Moon quotations from "Target: Agriculture?" *Farm Journal*, November 2001.

"Farms Present a Target Susceptible to Terror: Agricultural Resources Are Remote, Appear Easy to Attack," Scott Kilman, *Wall Street Journal*, December 26, 2001.

Much of the information about causes and effects of contaminated beef, of "mad cow" disease, and about the proposed Kevin's Law comes from the work of journalist Eric Schlosser, including *Fast Food Nation: The Dark Side of the All-American Meal*, Houghton Mifflin, New York, 2001; "Bad Meat: The Scandal of Our Food Safety System," *The Nation*, September 16, 2002; and "Order the Fish," *Vanity Fair*, November 2004.

Information about the 2002 ConAgra *E. coli* outbreak came from *The Denver Post*'s blanket coverage, including but not limited to "Parents, Children Relate Horror of Illness," David Migoya and Allison Sherry, July 12, 2002; "E. Coli Cases Grind on Butcher," Diane Carman, July 14, 2002; "Meat Inspection Flawed, Agency Says," David Migoya, July 17, 2002; "Critics Urge Reform of Beef-Recall Rules," David Migoya and Allison Sherry, July 21, 2002; "Allard's Food Inspection Votes Called into Question; USDA Recall Power Urged," Bill McAllister and Anne C. Mulkern, July 26, 2002; and "ConAgra Has List of Violations: Tainted Meat Found Dozens of Times at Greeley Facility," David Migoya, September 19, 2002.

Additional sources include these:

"Beef Processor's Parent No Stranger to Troubles," Greg Winter, *New York Times*, July 20, 2002.

"An Outbreak Waiting to Happen: Beef Inspection Failures Let in a Deadly Microbe," Joby Warrick, *Washington Post*, April 9, 2001.

"Lack of Oversight and Will Put Consumers at Risk," and "Inspector: Filth and Vermin Reported, Ignored," both by Oliver Prichard, *Philadelphia Inquirer*, May 18, 2003.

"Grandson's Death Turns Grove City Woman into Fighter for Safe Meat Laws," Karen Hoffman, *Pittsburgh Post-Gazette*, June 4, 2003.

"New Safety Rules Fail to Stop Tainted Meat," Melody Petersen and Christopher Drew, *New York Times*, October 10, 2003.

"Threat to Food Supply on Rise," Dennis O'Brien, *Baltimore Sun*, March 22, 2004.

THE AXIS OF WEEVIL

Information on the Nestlé boycott came from "Breaking the Rules, Stretching the Rules 2004: Evidence of Violations of the International Code of Marketing of Breastmilk Substitutes and Subsequent Resolutions," International Baby Food Action Network, 2004; and "Baby Milk Action: Nestlé's Inappropriate Language," *Corporate Watch*, Summer 1997, www.corporatewatch.org.uk/magazine/issue4/cw4cu.html.

Information about the 2002 USDA Agricultural Outlook Forum is from my own notes and from "The Future of Agricultural Biotechnology in World Trade: The Promise and Challenges," remarks delivered in Arlington, Va., by Alan P. Larson, Under Secretary for Economic, Business, and Agricultural Affairs, U.S. State Department, February 21, 2002, www.state.gov/e/rle/rm/2002/8447pf.htm.

"Monsanto Wins Gene Patent Case," Mike Lee, *Sacramento Bee*, May 22, 2004.

"Risking Corn, Risking Culture," Claire Hope Cummings, *Worldwatch*, November/December 2002.

"In Corn's Cradle, U.S. Imports Bury Family Farms," Tim Weiner, *New York Times*, February 26, 2002.

"Mexican Employment, Productivity, and Income a Decade After NAFTA," Sandra Polaski, director of the Trade, Equity and Development Project, Carnegie Endowment for International Peace, Washington, D.C., February 25, 2004.

Lighted-candle analogy from Richard A. Levins, *Willard Cochrane and the American Family Farm,* University of Nebraska Press, Lincoln, 2000.

WASTE LAND

Pertinent materials for this chapter include these items:

"The Prejudice Against Country People," Wendell Berry, essay in *Citizenship Papers,* Shoemaker and Hoard, Washington, D.C., 2005.

Swept Away: Chronic Hardship and Fresh Promise on the Rural Great Plains, Jon M. Bailey and Kim Preston, Center for Rural Affairs, Lyons, Nebraska, June 2003.

Home on the Range: A Century on the High Plains, James R. Dickenson, University Press of Kansas, Lawrence, 1995.

Donn Teske quotation from "Lay of the Land," *Insight,* April–May 2002.

"Stench Chokes Nebraska Meatpacking Town," Elliot Blair Smith, *USA Today,* February 14, 2000.

The Rap Sheet on Animal Factories: Convictions, Fines, Pollution Violations, and Regulatory Records on America's Animal Factories, Sierra Club, Washington, D.C., August 2002.

"A Flood of U.S. Corn Rips at Mexico," Michael Pollan, special to *Los Angeles Times,* April 23, 2004.

Drug Abuse in America—Rural Meth, Pilar Kraman, Trends Alert, Council of State Governments, Washington, D.C., March 2004.

"Once a Party Drug, Meth Moves into the Workplace," Daniel Costello, *Los Angeles Times,* September 13, 2004.

PHARAOH'S DREAM

Information for this chapter was drawn from the following sources:

2003 Annual Report, 2003 Fact Sheet, American Farmland Trust, Washington, D.C., www.farmlandinfo.org.

"A Landscape Assessment of the Catskill/Delaware Watersheds 1975–1998," Environmental Protection Agency, www.epa.gov/nerl.

"Seeking Safety, Manhattan Firms Are Scattering," Charles V. Bagli, *New York Times*, January 29, 2002.

"Destination #2: Enterprise Boll Weevil Monument," Rick Harmon, *Montgomery* (Alabama) *Advertiser*, January 24, 2002.

AFTERWORD: WASPS AND FINCHES

Sources of inspiration and useful information for this section include these items:

Rocks of Ages: Science and Religion in the Fullness of Life, Stephen Jay Gould, Library of Contemporary Thought, Ballantine Publishing, New York, 1999.

Agriculture Department statement from *Trends in U.S. Agriculture*, U.S. Department of Agriculture, National Agricultural Statistics Service, www.usda.gov/nass/pubs/trends.

Lords of the Harvest: Biotech, Big Money, and the Future of Food, Daniel Charles, Perseus Publishing, 2001.

Good examples of independent farming are from "The 2004 Patrick Madden Awards for Sustainable Agriculture," Sustainable Agriculture Research and Education Program, USDA, Washington, D.C., 2004.

Index

PublicAffairs is a publishing house founded in 1997. It is a tribute to the standards, values, and flair of three persons who have served as mentors to countless reporters, writers, editors, and book people of all kinds, including me.

I. F. Stone, proprietor of *I. F. Stone's Weekly,* combined a commitment to the First Amendment with entrepreneurial zeal and reporting skill and became one of the great independent journalists in American history. At the age of eighty, Izzy published *The Trial of Socrates,* which was a national bestseller. He wrote the book after he taught himself ancient Greek.

Benjamin C. Bradlee was for nearly thirty years the charismatic editorial leader of *The Washington Post.* It was Ben who gave the *Post* the range and courage to pursue such historic issues as Watergate. He supported his reporters with a tenacity that made them fearless, and it is no accident that so many became authors of influential, best-selling books.

Robert L. Bernstein, the chief executive of Random House for more than a quarter century, guided one of the nation's premier publishing houses. Bob was personally responsible for many books of political dissent and argument that challenged tyranny around the globe. He is also the founder and was the longtime chair of Human Rights Watch, one of the most respected human rights organizations in the world.

.   .   .

For fifty years, the banner of Public Affairs Press was carried by its owner Morris B. Schnapper, who published Gandhi, Nasser, Toynbee, Truman, and about 1,500 other authors. In 1983 Schnapper was described by *The Washington Post* as "a redoubtable gadfly." His legacy will endure in the books to come.

Peter Osnos, *Publisher*